WILD NEW ZEALAND

WILD NEW ZEALAND

Photographs by **GERALD CUBITT**

Text by **LES MOLLOY**

Consultants: SUE MILLER and BRIAN COFFEY

In collaboration with the New Zealand Department of Conservation

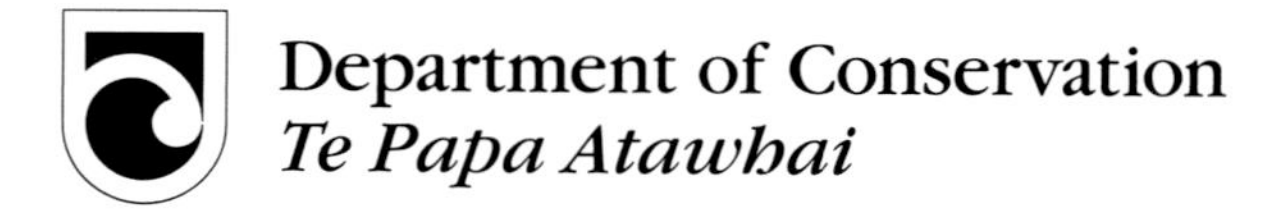

The MIT Press
Cambridge, Massachusetts

First MIT Press edition, 1994

Library of Congress Cataloging-in-Publication Data

Wild New Zealand / photographs by Gerald Cubitt : text by Les Molloy in collaboration with the New Zealand Department of Conservation : consultants, Sue Miller and Brian Coffey. – 1st ed.
p. cm.
Includes bibliographical references and index.
ISBN 0-262-13304-0
1. Natural history – New Zealand. 2. Zoology – New Zealand. 3. Nature conservation – New Zealand. I. Molloy, Les. II. New Zealand. Dept of Conservation. III. World Wide Fund for Nature New Zealand.
QH197.5.W534 1994
508.93–dc20

94-27175
CIP

Publishing Director: Charlotte Parry-Crooke
Commissioning Editor/Project Manager: Ann Baggaley
Editors: Paul Barnett, John Stidworthy
Designer: Behram Kapadia
Cartography: Julian Baker
Index: Paul Barnett

Reproduction by Typongraph, Verona, Italy
Printed and bound in Singapore by Kyodo Printing Co (Pte) Ltd

CONTENTS

Photographic Acknowledgements

The publishers and photographer extend their thanks to the following people who kindly loaned their photographs for inclusion in this book. All the photographs in the book, with the exception of those listed below, were taken by Gerald Cubitt.

Brian Coffey: page128 (bottom right); page 181 (left).

Rob Greenaway: page 52 (bottom).

Ken Grange: page 159 (top left); page 165 (bottom, all three subjects).

E. Humphries: page 83 (centre).

Sue Miller: page 51; page 52 (top left and top right); page 53 (both subjects).

Les Molloy: page 21 (bottom); page 23; page 25; page 26 (bottom); page 32 (bottom); page 34 (both subjects); page 35 (top); page 43 (right); page 45; page 90 (bottom); page 116 (bottom); page 117 (bottom); page 118 (top); page 132/3; page 140; page 142 (bottom left); page 145 (top); page 148 (top and bottom); page 152 (top left and top right); page 173; page 174 (bottom left); page 181 (top right and bottom right).

Rod Morris: page 18 (left).

New Zealand Department of Conservation: page 118 (bottom left); page 119 (bottom left); page 126 (centre right); page 130 (centre right); page 134; page 151 (upper left and lower left); page 179 (bottom left); page 200 (centre left and centre right; bottom left and bottom right).

K. Westerhov: page 133 (centre and bottom).

Illustrations appearing in the preliminary pages are as follows:

HALF-TITLE: Takahe (*Porphyrio mantelli*), Tiritiri Matangi Island.
FRONTISPIECE: Wild river falls, Central North Island.
PAGE 5: Kakabeak (*Clianthus puniceus*), Urewera National Park.
PAGE 7: Elephant Seal (*Mirounga leonina*) pup, Auckland Islands.
PAGE 10: Marlborough Sounds viewed from Maud Island.

Acknowledgements

The author, photographer and publishers wish to express their thanks to the following for providing much generous and valuable assistance during the preparation of this book:

World Wide Fund for Nature New Zealand (WWF-NZ)
Chris Laidlaw, Executive Director • Sue Miller, former Conservation Scientist (in association with Brian Coffey, Brian T. Coffey and Associates Ltd, Environmental Consultants) • Dave Boardman, former Marketing Director

New Zealand Department of Conservation
Sonia Frimmel, Visitor Services Division
Hugh Best, Dr Rod Hay, Dr Hugh Robertson and Carol West of the Science and Research Division • Dr Don Merton and Alan Saunders of the Threatened Species Unit • Dr Wren Green, Shona Mackay and Ferne McKenzie of the Planning and External Agencies Division

In the field, advice and support was offered by:
Mike Aviss • Shaarina Boyd • Derek Brown • Lindsay Chatterton • Gideon Climo • David Crouchley • Daryl Eason • Margaret Edridge • Raewyn Empson • Martin Gembitsky • Pearl Hewson • Ken Hunt • Paul Jansen • Graeme Loh • John Lyall• John Lythgoe • Dennis McDonnell • Steve McManus • Alan Munn • Boyd Parker • Richard Parrish • Phil Thomson • Ray Pierce • Gretchen Rasch • Christine Reed • Ray Scrimgeour (also Doug and Suzanne at Te Kuiti) • Malcolm Smith • Chris Smuts-Kennedy • Kay Stark • Ray and Barbara Walter • Bruce Watson

Particular thanks and appreciation are extended to:
Air New Zealand
Mount Cook Airlines • New Zealand Tourism Board • Budget Rent-a-Car • Southern Heritage Tours (Subantarctic and Fiordland) • South Pacific Hotel Corporation (Auckland Parkroyal; Christchurch Parkroyal; Queenstown Parkroyal; Wellington Parkroyal; Christchurch Airport Travelodge; Hermitage Hotel, Mount Cook; Te Anau Travelodge; The Milford Guided Walk and the Routeburn Guided Walk)

Abel Tasman National Park Enterprises • Brian Karl, Landcare Research • Dave Patterson, New Zealand Outdoor Adventures • Wellington Photographic Supplies • Eric Fox, Otorohanga Kiwi House • Fiordland Travel • James Heyward, New Zealand Nomad Safaris • John and Denise Sutherland, Chatham Lodge, Chatham Islands • Monarch Otago Harbour Cruises • Pacifica Lodges and Inns • Phill Cooney, Mount Cook Airline Ski Plane Division • Robert Fleming, Vulcan Helicopters • Ross Leger, Kapiti Marine Charters • Scott Clarke, Penguin Place Wildlife Area/McGrouthers Farm • Steve Collins, Tarawera Helicopters • Steve Murphy, C.R. Kennedy Ltd (Hasselblad services) • Tom and Liz True, Mercury Bay Beachfront Resort • White Heron Sanctuary Tours

Janet Cubitt

Dr Ian Atkinson • Brian Bell • Susan Buckland • David Crockett • Dr Charles Daugherty • Paul and Phyllis Every • Bernard Goetz • Megan Griffiths • Peter and Susie Lindauer • Alan Lorking • Dr Rob Mattlin • Folkert Nieuwland • Noel and Val Parsons • Brett Robertson • Rodney and Shirley Russ • Lorna Simpkin • Nadine Stander • Gavin Woodward

Alan McEldowney, South Pacific Books • Dr Michael Green, World Conservation Monitoring Centre, Cambridge, England • Bing Lucas, former Chairman, IUCN Commission for National Parks and Protected Areas, Wellington

Special acknowledgements go to:
Ian Higgins, former Executive Director, WWF-NZ, for his involvement and support for this project from its conception; also Professor David Bellamy for his kind contribution

PREFACE

About a thousand years ago the ancestors of the Maori people were just discovering New Zealand – their Aotearoa, the 'Land of the Long White Cloud'. It was the last large piece of the temperate and sub-tropical world to be colonized by men and ground-living mammals.

Up until that time a unique flora, derived from that of ancient Gondwanaland, had flourished almost undisturbed in a setting of young mountains, active volcanoes and glaciers. Flightless birds both large and small, such as the Moa, lived without threat along with a diverse avifauna which still boasts Kiwis, Takahes, Kakapos, Wrybills, Penguins, the only alpine parrot and some of the world's best songsters. The coming of humans and the introduction of feral animals wrought great changes in the natural balance of this hitherto untouched land, and for some species the inevitable outcome was extinction.

Today, New Zealand's people, both Maori and Pakeha, understand the fragility of their heritage and are doing something about it. With the dedication and expertise of such bodies as the Department of Conservation, the Maruia Society, the Native Forest Restoration Trust, the Royal Forest and Bird Protection Society, WWF and many others, the country's biodiversity is being put back into working order.

To me, this is still the most exciting natural history location on earth, a wonderland of geology, botany and zoology, spectacular in every dimension. From the serpentine cliffs of Cape Reinga in the north to the enormous World Heritage Site of Fiordland and beyond to the Subantarctic Islands, it is a breathtakingly beautiful place – a workshop of evolution and conservation in action.

David Bellamy

David Bellamy
The Conservation Foundation

INTRODUCTION

New Zealand is an enigma. It contains some of the world's wildest places, like Milford Sound, the Southern Alps, Mount Taranaki (Egmont) and the rainforests of Westland. These well known images attest to the natural beauty of a remote land, relatively free from the problems of over-population, ethnic conflict and industrial pollution that blight less fortunate nations. But New Zealand is also a modern developed society, with highly efficient agricultural and forestry industries; a land so transformed during 150 years of European settlement that most of the lowlands and much of the coastlands are anything but wild any more. Indeed, many visitors are just as attracted to the serene pastoral landscapes dotted with sheep and poplar trees, paddocks neatly dissected by wire fences and shelterbelts, as they are to wilderness. Many tourists seek out the rural 'clean and green' idyll of bountiful kiwifruit and grape vineyards, apple orchards, deer farms and pine plantations. Yet all these agricultural crops and herds have been introduced from Northern Hemisphere temperate lands, while virtually none of New Zealand's indigenous land plants and animals have been domesticated or are any longer used in a commercially sustainable way. As a consequence, there is a curious dichotomy about modern New Zealand, with two parallel landscape systems co-existing: the natural world of the smaller islands and mainland 'back-country' wilderness, and the highly ordered world of the rural countryside.

Other aspects of the country's natural history are puzzling. For example, New Zealand is sometimes referred to as 'the ancient islands' because of the antiquity of its plants, insects, amphibians, freshwater fish, reptiles and birds. Many of these were originally inhabitants of the southern supercontinent of Gondwana, and weathered an incredible journey through time on what the noted British naturalist Dr David Bellamy has called 'Moa's Ark'. Over 80 million years ago the continental fragment that was to become New Zealand broke away from Australia-Antarctica and headed off into the Southern Ocean. Today, it is difficult to find any landforms in New Zealand older than a mere 14,000 years, so overwhelming has been the effect on the land of enormous geological and climatic forces – earthquake and mountain uplift, volcanic eruption and ice-age glaciation, river down-cutting and sea-level rise. In addition, the insidious agents of erosion have both worn away and rejuvenated the land through the regular cycles of sun and cloud, rain and wind. Yet somehow much of New Zealand's unique biota has survived the turbulent environmental changes of that 80 million years of geological history.

New Zealand has many other surprises. These islands, comprising one of the most isolated archipelagoes on Earth, could on the basis of their geology and biota be called the world's smallest continent. They are South Pacific islands with many plants of tropical origin, yet around some of these plants there are glaciers descending almost to sea level. No snakes or, save two species of bats, native land mammals can be found on this ancient ark, but evolution has allowed some groups of insects and birds to fill many ecological niches normally occupied by mammals. Furthermore, New Zealand's youthful mountains, damp evergreen forest (the Bush) and grey rocks stand in sharp contrast to the landscape of her nearest neighbour – the old, dry, flat, red continent of Australia.

Geography, Landforms and Climate

The New Zealand archipelago consists of about 700 islands with a total area of about 270,000 square kilometres (104,250 square miles). Two of the islands are much larger than all the others, together making up more than 98 per cent of the country: the South Island (56.1 per cent) and the North Island (42.3 per cent). Their rather prosaic English names have survived through convenience, but two of their traditional Maori names are much more descriptive, evoking the shape of each with extraordinary accuracy: Te Waka a Maui ('The Canoe of Maui') for the South Island and Te Ika a Maui ('The Fish of Maui') for the North Island.

Although all the other islands together constitute only 1.6 per cent of the land area, they add immeasurably to the geographical and biological diversity of the country. In particular, three outlying island groups markedly extend the territorial waters of New Zealand and determine a maritime Exclusive Economic Zone that is one of the most extensive possessed by any nation. The northernmost island is Raoul Island, in the subtropical Kermadec group 1,000 kilometres (620 miles) to the north-east of the North Island, at latitude 29 degrees South; the southernmost is one of the subantarctic islands, Campbell Island, 700 kilometres (435 miles) south of the southern coast of the South Island, at latitude 52.5 degrees South. In all, therefore, New Zealand has an impressive latitudinal range of 23.5 degrees. The third of these significant island groups, the Chatham Islands, lies 800 kilometres (500 miles) to the east of the South Island on the Chatham Rise, an eastward extension of the New Zealand continental shelf.

Three physical factors dominate the environment of New Zealand: the country is maritime, extremely mountainous and, especially in the North Island, subject to the pervasive influence of volcanoes.

Maritime influences pervade not only the 700 smaller islands but also the mainland (i.e., the North and South Islands), for no part of the country is more than 130 kilometres (80 miles) from the sea. The coastline is very long – 15,000 kilometres (9,300 miles) – because of the large number of small islands and because parts of the mainland, especially the Hauraki Gulf, Marlborough Sounds, Banks Peninsula and Fiordland, have a very intricate coastal pattern.

As a result of its location in southern-temperate latitudes, New Zealand is in the path of the West Wind Drift (also known as the Tasman Current), the wide oceanic surface current that encircles the Southern Hemisphere from west to east. The Drift scours the western coasts and surges north around Cape Reinga and south around Stewart Island and through Foveaux Strait; a particularly powerful current squeezes through the narrow Cook Strait, between the North and South Islands, giving rise to tidal speeds up to nine kilometres (5.6 miles) per hour. In general, therefore, currents tend to be clockwise around the northern half of the

Maritime influences pervade most of New Zealand, for no part is more than 130 kilometres (80 miles) from the sea and the 15,000-kilometre (9,300-mile) coastline includes many estuaries and islands.

country and anticlockwise in the south. The warm waters of the West Wind Drift – warm because of their subtropical origin – come into contact with cooler, less saline waters of subantarctic origin around the southern end of the South Island. This marine transitional zone, the Subtropical Convergence, is a major feature of the southern oceans and is responsible for sharp differences in marine life across its boundaries. Marine plants and animals that can tolerate variations in temperature and salinity tend to inhabit the convergence or the shallower coastal waters of the continental shelf.

Mountains and hills dominate the New Zealand skyline. This is in a way ironic. The country's English name is an anglicization of 'Niew Zeeland', one of the names conferred by the Dutch navigators who became aware of this southern land in the mid-1600s, yet Zeeland itself is a low-lying coastal province of the Netherlands, completely lacking in mountains.

In New Zealand, by contrast, the proximity of the mountains

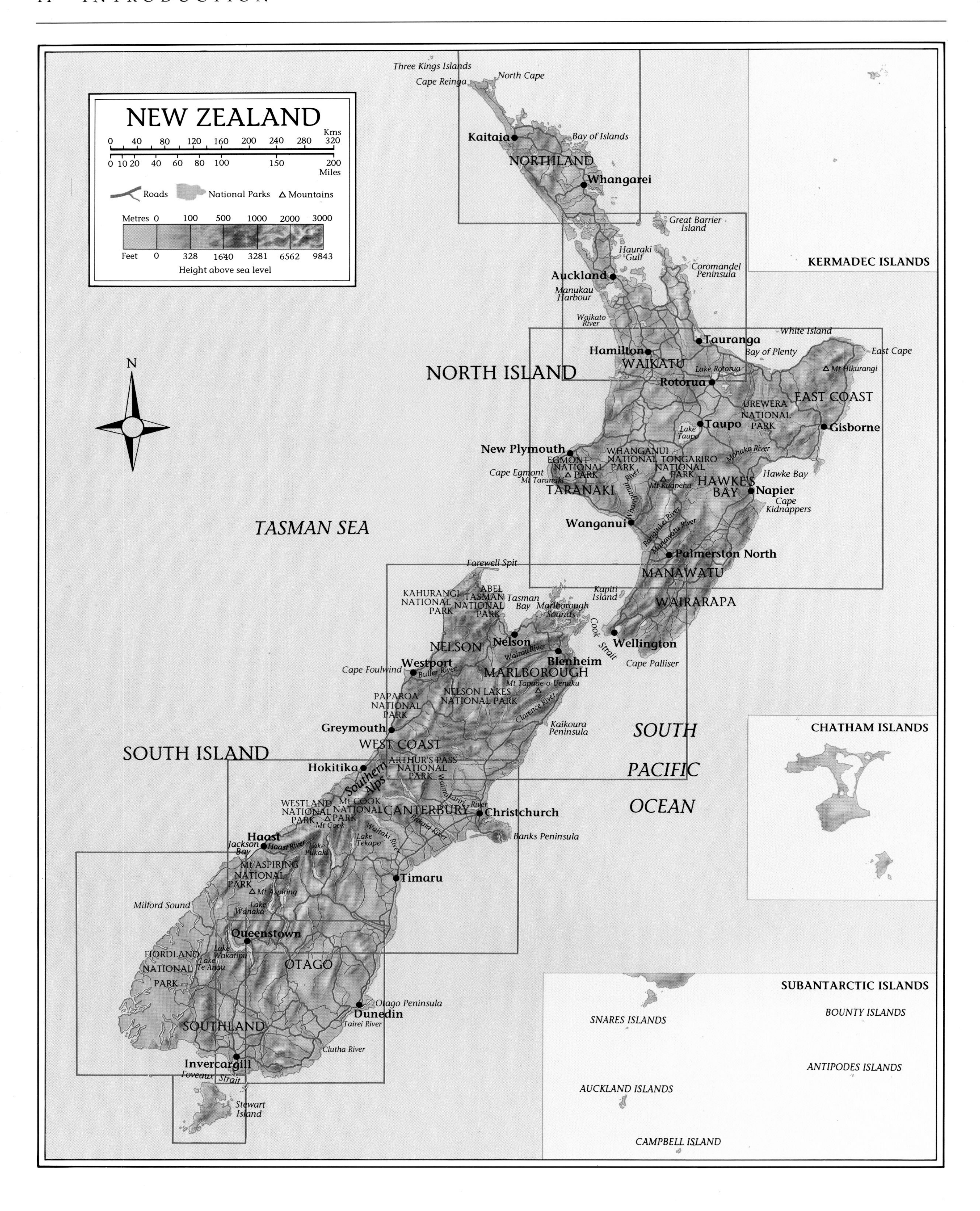
NEW ZEALAND
Kms
Miles
Roads
National Parks
Mountains
Metres 0 100 500 1000 2000 3000
Feet 0 328 1640 3281 6562 9843
Height above sea level
N
Three Kings Islands
Cape Reinga
North Cape
Kaitaia
Bay of Islands
NORTHLAND
Whangarei
Great Barrier Island
Hauraki Gulf
Coromandel Peninsula
Auckland
Manukau Harbour
Waikato River
KERMADEC ISLANDS
White Island
Tauranga
Bay of Plenty
East Cape
Hamilton
WAIKATO
NORTH ISLAND
Lake Rotorua
Rotorua
Mt Hikurangi
UREWERA NATIONAL PARK
EAST COAST
Taupo
Lake Taupo
Gisborne
New Plymouth
EGMONT NATIONAL PARK
WHANGANUI NATIONAL PARK
TONGARIRO NATIONAL PARK
Mohaka River
Cape Egmont
Mt Taranaki
Mt Ruapehu
Hawke Bay
TARANAKI
HAWKES BAY
Napier
Cape Kidnappers
Whanganui River
TASMAN SEA
Wanganui
Rangitikei River
Manawatu River
Palmerston North
MANAWATU
Farewell Spit
KAHURANGI NATIONAL PARK
ABEL TASMAN NATIONAL PARK
Tasman Bay
Marlborough Sounds
Kapiti Island
WAIRARAPA
Cook Strait
NELSON
Nelson
Wairau River
Blenheim
Wellington
Cape Palliser
Westport
Cape Foulwind
Buller River
MARLBOROUGH
Mt Tapuae-o-Uenuku
PAPAROA NATIONAL PARK
NELSON LAKES NATIONAL PARK
Clarence River
Kaikoura Peninsula
Greymouth
SOUTH PACIFIC OCEAN
CHATHAM ISLANDS
WEST COAST
SOUTH ISLAND
Hokitika
ARTHUR'S PASS NATIONAL PARK
Southern Alps
Waimakariri River
WESTLAND NATIONAL PARK
Mt COOK NATIONAL PARK
Mt Cook
CANTERBURY
Christchurch
Rakaia River
Banks Peninsula
Haast
Jackson Bay
Haast River
Lake Pukaki
Lake Tekapo
Waitaki River
Mt ASPIRING NATIONAL PARK
Mt Aspiring
Timaru
Milford Sound
Lake Wanaka
Queenstown
Lake Wakatipu
FIORDLAND NATIONAL PARK
Lake Te Anau
OTAGO
SUBANTARCTIC ISLANDS
Otago Peninsula
Dunedin
Tairei River
SNARES ISLANDS
BOUNTY ISLANDS
SOUTHLAND
Clutha River
Invercargill
Foveaux Strait
ANTIPODES ISLANDS
AUCKLAND ISLANDS
Stewart Island
CAMPBELL ISLAND

RIGHT Miranda, in the Firth of Thames, is one of many estuaries along the coastlines of Northland and Auckland that become winter havens for huge flocks of migratory birds.

BELOW Mountains dominate the South Island's skylines. At 3,754 metres (12,316 feet), Mount Cook (also known by its Maori name, Aoraki) is the highest of many peaks above 3,000 metres (9,800 feet) in the Southern Alps.

Looking northwards across the summit of Mount Ruapehu (with its crater lake) to Mount Ngauruhoe. The most active of New Zealand's volcanoes lie in the central North Island.

to the coastlines is one of the geographical features that makes the landscape so dramatic; nowhere is this more evident than in the Wellington, Kaikoura, West Coast and Fiordland regions. Overall, 60 per cent of the country is higher than 300 metres (985 feet) and 70 per cent of it is either hilly or steep. Most of the mountainous country is in the South Island, where a chain of mountains extends down the entire length of the island, running north-east to south-west from the Bryant and Richmond ranges (between Nelson and Marlborough) through the Spenser Mountains and the Southern Alps to the mountain wilderness of Fiordland. In the central part of the Southern Alps the mountains are 3,000-3,700 metres (9,840-12,140 feet) above sea level, with Mount Cook (known to the Maori people as Aoraki) the highest peak in the country, at 3,754 metres (12,316 feet). The North Island is more hilly than mountainous in character, although mountain ranges extend in a line from Cook Strait to East Cape and volcanic mountains are prominent landmarks in Taranaki and the centre of the island. Earthquakes are common along fault-lines in these mountainous areas, where ranges are being rapidly uplifted, at rates up to one metre (40 inches) per century.

Volcanoes and geothermal phenomena are widespread throughout the archipelago. The best-known volcanic landmarks are in the western and central parts of the North Island: Mount Taranaki (Egmont), the active volcanoes of Tongariro National Park, Lake Taupo, the Rotorua geyser-fields, Mount Tarawera and White Island. These volcanoes are both andesitic and rhyolitic in composition and have a history of extremely violent eruptions, particularly the eruptions from the calderas now filled by Lake Taupo and Lake Rotorua.

The Coromandel Peninsula, Auckland and the Bay of Islands hinterland are also volcanic, the two latter localities being well known for their many small, steep-sided basaltic lava and scoria (cinder) cones. Rangitoto Island, on Auckland's eastern skyline, is the youngest and most beautiful of these volcanoes. The smooth sweep of the eastern coastline of the South Island is disrupted by highly eroded remnants of two groups of very old volcanoes: Banks Peninsula and Otago Peninsula. Furthermore, many of the spectacular offshore and outlying islands are former volcanoes, notable examples being Mayor Island (Tuhua) and Little Barrier Island (Hauturu). Volcanoes mark even the extremities of New Zealand: active in the case of Raoul and Curtis islands, in the Kermadec group, and extinct in that of Campbell Island, far to the south of Stewart Island.

To a large extent New Zealand's climate is determined by its maritime setting and its mountainous character. Essentially the country is a wall of mountains standing in the path of the moisture-laden westerly winds that circle the globe constantly in these southern-temperate latitudes. As this maritime air is forced up over the mountains it cools rapidly, and the moisture condenses as cloud and rain. This is particularly so along the Southern Alps and the rest of the main divide of the South Island, a barrier 750 kilometres (470 miles) long and over 1,500 metres (4,920 feet) tall. This unbroken mountain wall gives the West Coast its super-humid climate, the life-source of its rainforests, wetlands and glaciers. At sea level in Milford Sound the average annual rainfall is over 6,000 millimetres (236 inches), while high on the flanks of the Southern Alps above Whataroa is one of the wettest places in the world with average rainfall of a stupendous 15,000 millimetres (590 inches) per year. At the other extreme, the basins in the lee (rain-shadow) of the Southern Alps are the driest places in the country: around Alexandra, in Central Otago, the annual rainfall is as low as 350 millimetres (under 14 inches), summers are hot and dry, winters frosty and dry and the landscape semi-arid. These basins are the only parts of New Zealand where the maritime influence weakens and the climate is more continental.

It is often said that the most striking feature of the New Zealand landscape is its diverse scenery, reflecting the sharp changes in rock-type, landforms, vegetation and climate that occur across the country. The diversity of scenery normally to be found on a whole continent is squeezed into the two relatively small main islands. For example, in a single day's car journey it is possible to enjoy the basalt cliffs and cool maritime air of Otago Peninsula, the loess-mantled tussocklands and montane bogs of the East Otago uplands, the craggy schist tors of hot, dry Central Otago, the ice-gouged troughs filled by the great southern glacial lakes Hawea and Wanaka, the beech forests and braided rivers of the Makarora and Haast valleys, the herb-fields and fell-fields above Haast Pass, the scarp of the Alpine Fault (marking the active boundary between the two crustal plates New Zealand straddles) and the wet Haast coastal plain, dotted with Kahikatea swamp forest and wind-shorn Rimu hugging old dunes, and with the surf of the Tasman Sea beyond.

Ancient Origins of New Zealand's Biota

> The biotas of New Zealand and Madagascar are the closest we shall ever come to observing the products of continental evolution in island-like isolation, unless we discover higher life on another planet.
>
> Jared M. Diamond, 'New Zealand as an Archipelago' (1989)

The story of the origins of New Zealand and its unusual plants and animals is a long one – and one that gets longer and more fascinating as scientists gradually unravel the clues locked away in fossil-containing rocks, ocean-sediment cores, peat bogs, volcanic tephra deposits and the bones of extinct animals long ago washed into limestone caves.

Four overriding environmental factors have combined to account for the uniqueness of New Zealand's biota. Two of these are the maritime temperate climate and the mountainous topography. The third factor, shared by a number of oceanic island groups, is the extreme isolation from other major landmasses. In New Zealand's case this isolation has obtained for over 80 million years, and, while other isolated volcanic island groups, like the Hawaiian and Galápagos islands, are similarly mountainous and contain some very unusual plant and animal life, they differ markedly from New Zealand because their volcanoes are relatively young and because they have no continental character: their lifeforms must have rafted or been blown across the ocean to their shores.

This continental character of the archipelago and much of its biota is the fourth and crucial factor. At a certain point in the planet's evolutionary history, about 80 million years ago, a combination of geological forces caused a fragment to break away from the ancient southern supercontinent Gondwana. That fragment eventually became the group of islands we call New Zealand; it was reached by human beings a mere 1,000 years ago.

While it is truly remarkable that many of the animals and plants which lived in the warmer climes of Gondwana 100 million years ago survived the journey into temperate isolation, we should not imagine that this Gondwana cargo went into some sort of suspended animation; rather, during the long isolation from the other continents, it continued to evolve in response to the changing physical environment. Furthermore, the ancestral stock was added to from time to time – especially by plants and birds carried along on the westerly winds and the West Wind Drift. Biologists describe the New Zealand biota as being depauperate (i.e., lacking in major groups of plants and animals) but very rich in lifeforms which have developed from the limited parent stock or have been, since the separation, elsewhere eliminated by superior (in an evolutionary sense) organisms. In essence, then, while there is great diversity in some biota – such as seabirds and alpine plants – the attraction of most of New Zealand's wildlife stems from its curiosities and its antiquity.

The timing of the separation of the ancestral New Zealand landmass meant that the biological cargo aboard the ark was representative of the Cretaceous period, the latter part of the Age of the Dinosaurs. Recent exciting fossil discoveries in Hawke's Bay tell us that a range of dinosaurs – pterosaurs and marine plesiosaurs, sauropods and carnosaurs – did set out on the voyage. Like the rest of the world's dinosaurs, they failed to survive the cataclysm at the end of the Cretaceous, some 65 million years ago.

Some animals did not 'catch the boat' 80 million years ago for the very simple reason that they were not then around in significant numbers. The most notable omissions from New Zealand's biota are land mammals, snakes and many of the families of flowering plants (angiosperms); these established their dominance only at the end of the Cretaceous, after the boat had gone. The host of Palaeozoic and Mesozoic lifeforms for which the New Zealand ark provided refuge were largely extinguished in the other Gondwana continents during the succeeding Tertiary period, as more highly evolved plants and animals became dominant.

Whole orders and families of higher animals are found only in New Zealand, in particular Tuatara (the sole representative of one of the four orders of reptiles), Moas (now extinct) and Kiwis. New Zealand's lizards are also endemic (that is, unique to New Zealand, often to regions or localities within the country), and for its size New Zealand has a very diverse range of skinks and geckos (60 species). Tuatara can live for more than 70 years and Duvaucel's Gecko for 35 years – the greatest lifespan recorded for a lizard. Geckos are considered quite ancient lizards whose centre of origin was in this southern continent. The four species of native frog all belong to the very primitive, endemic Leiopelmatidae family, whose lineage also reaches back to Gondwana.

Lower forms of animal life are often overlooked in favour of a country's more attractive or spectacular animals (such as, in New Zealand's case, Kiwi, Kakapo, Takahe and Kea). However, the country's smaller and lesser-known animals are of great interest to scientists because of their antiquity, long isolation and very high level of endemism (about 95 per cent). Many spiders, bristletails, moths and dragonflies have features that elsewhere in the world can be found only in fossils. Caddisflies, stoneflies and mayflies – all insects that frequent cold running water – comprise another group with strong links to Gondwana. A feature of the beetles and other insects is their flightlessness and their large size, attributes of invertebrates living in cold, windy climates where there are no mammalian predators. Some of the most spectacular insects are the large, flightless weevils, such as the giant 'giraffe weevils'.

Other endemic and often primitive lower-animal groups are the flatworms, giant earthworms, giant land-snails and (see below) *Peripatus*. Nearly 200 species of native earthworms are known,

Hochstetter's Land Snail (*Powelliphanta hochstetteri*), found high in the mountains of Nelson, is typical of New Zealand's rich fauna of endemic, ancient snails which survived in the absence of mammalian predators.

The Poor Knights Giant Weta (*Deinacrida fallai*) is one of 70 species of this frightening-looking group of endemic insects. Wetas have evolved to fill many of the niches elsewhere occupied by mice.

most of them restricted to remnant forest habitats because economically important introduced earthworms have displaced them from soils in gardens and pastures. Some of the native earthworms can exceed one metre (40 inches) in length and one centimetre (0.4 inches) in diameter, and are related to similar species found in isolated habitats in other Gondwana fragments like Australia and South America. The land-snail fauna of New Zealand is one of the richest in the world, reflecting the primaeval wet-forest environment – and, probably, the lack of mammalian consumers.

Like the earthworms, most of the over 1,000 known species of snail inhabit native forest, especially the leaf litter and humus layers of the forest floor. They, too, have been replaced in gardens and agricultural lands by the ubiquitous European Snail, a horticultural pest. Some of these native land-snail genera are large and colourful, and most are under serious threat from introduced mammals (rats, pigs and hedgehogs) and birds (e.g., thrushes and blackbirds). A feature of the land-snails is their restricted geographical distribution, often reflecting land connections that existed earlier during the evolution of the country's shape through uplift, glaciation and erosion. Two groups are restricted to remote parts of Northland or to northern islands: the Flax Snails (*Placostylus*), with their distinctive spiral shells up to 12 centimetres (nearly five inches) in length, and the Kauri Snails (*Paryphanta*), which have the ability to climb Kauri trees and travel hundreds of metres while foraging for worms at night. The giant land-snails belong to the Powelliphanta group, members of the Rhytididae, a widespread Gondwana family of snails considered the world's oldest family of carnivorous land-snails. The Powelliphanta have coiled shells that are often ten centimetres (four inches) in diameter, and their geographical isolation – often high in the mountain ranges of the southern North Island and north-west Nelson – has given rise to species with shells of many different attractive patterns and colours. Like so many among those of New Zealand's ancient fauna that display gigantism, they can be long-lived (up to 40 years).

Peripatus is one of the most curious of the ancient inhabitants of New Zealand's forests. It can be traced back to the middle of the Cambrian period (550 million years ago), and has long fascinated zoologists because it is neither worm, centipede nor caterpillar, but has affinities with all three. It is now considered the 'missing link' between the segmented worms, like earthworms, and the arthropods (centipedes, spiders and insects). Like Powelliphanta land-snails, these carnivores cannot withstand desiccation and hunt the moist forest floor at night. They entrap and consume their prey – insect larvae and soft-bodied arthropods – in a rather gruesome way, ensnaring the victim in a sticky slime ejected from the mouth, then cutting a hole in the side of the body and injecting a powerful saliva, which dissolves the body contents so that the unfortunate prey can then be sucked dry.

The insects of one group stand out for their rather striking – indeed, frightening – appearance and for their antiquity. Commonly known by their Maori name, weta, they are members of the large insect order Orthoptera (which also includes grasshoppers, locusts and crickets). In many ways the weta is New Zealand's most interesting animal (and, while every New Zealander has seen a weta, very few will ever see their national symbol, the Kiwi, in the wild). There are more than 70 species, all endemic. Many believe the giant wetas of New Zealand to be among the most ancient and unchanged members of the Orthoptera, dating back nearly 200 million years in the Gondwana fossil record. In an evolutionary response to the lack of mammalian foragers and predators, these dinosaurs of the insect world have developed heavy bodies – weighing as much as 70 grams (2½ ounces) – and resemble small rodents like mice and rats in their diet and in their strictly nocturnal activity. Varieties of the ubiquitous weta are as much at home in domestic woodpiles, basements and gardens as in their natural forest, cave and alpine habitats.

Like wetas, Kiwis have a very ancient lineage. They have evolved into flightless, nocturnal birds with many attributes more typical of mammals. If we disregard the long beak (used to probe for soil animals), Kiwis even look like mammals, with their shaggy feathers giving a superficial appearance of fur. Like a cat, Kiwis have sensory whiskers, and like most mammals (and unlike birds) they have an acute sense of smell; they are the only known birds to have external nostrils (at the end of the beak!). They burrow in the ground like moles, exude an earthy musty smell and may mark out their territory by means of their strongly smelling burrows. Their body temperature is lower than that of other birds and closer to that of mammals. Indeed, this is the most 'unbirdlike' of birds. Certainly it is distinctive – worthy of its status as a national symbol for modern New Zealanders, who refer to themselves colloquially as 'Kiwis'.

The Kiwi is only the best-known of New Zealand's large and flightless birds: there are (and were) others. Three which still exist

A Takahe chick being fed by hand-puppet 'parents' in a captive-rearing recovery programme at Burwood Bush Scientific Reserve. A small Takahe population (under 200) lives in the wilds of Fiordland National Park.

in the wild are the Takahe and Weka (both rails) and the Kakapo (a parrot). The Takahe is the largest of New Zealand's remaining flightless birds; it occupies a unique position in the country's avian folklore because it was presumed extinct for 50 years until 'rediscovered' in a remote Fiordland mountain valley in 1948. Although fewer than 200 birds are thought to remain in the Murchison and Stuart mountains of Fiordland, it has been bred successfully in captivity. The Weka is an endemic rail that is very obvious in the wild because it is not only large and flightless but very inquisitive. This bold and fearless bird, so measured in its deliberate steps and tail-flicking, is capable of killing rats and young seabirds. Even so, three of its four subspecies are in steady decline: the North Island, Eastern (Buff) and Stewart Island wekas.

The flightless, nocturnal Kakapo is the world's largest and rarest parrot. With its large eye-discs it resembles an owl, and it has a strong musk-like smell like many mammals.

The Kakapo is the world's rarest and largest parrot; indeed, it is one of the rarest of all birds, and also one of the most interesting. With its large eye-discs it looks somewhat like an owl, and also like an owl it is nocturnal – hence one of its early European names, the 'owl-parrot'. It has a strong, musk-like smell, like many mammals; it is solitary, and the male is the only member of the Parrot family known to perform nightly lek displays during the breeding season. (Lek displays are characteristic of male birds which try to attract potential female mates to their 'arenas' by a combination of calls and displays.) The Kakapo male develops an intricate 'track and bowl' system for the courtship, but seems to initiate its calling ('booming') only in seasons when there is a plentiful supply of high-protein food. The low-frequency booming of male Kakapos during these lonely vigils can be heard up to five kilometres (three miles) away! The abandonment of flight has led to a considerable increase in the Kakapo's body size: they can be up to 65 centimetres (two feet) long.

Other notable large flightless birds are now extinct. The most famous is the Moa, but three others were at least a metre (40 inches) tall: a goose (*Cnemiornis*), the 'giant penguin' (*Pachydyptes ponderosus*), which was about 1.6 metres (over five feet) tall, and a rail-like bird (*Aptornis otidiformis*), popularly known as the Adze-bill because of its impressive beak.

Like the Kiwi, the Moa belonged to the Ratite family, which includes a number of very large birds that seem to have opted to give up the power of flight and instead develop powerful legs for running away from (or defending themselves from) predators. Other Gondwana fragments took with them their own version of the Moa – the Ostrich of Africa, the Elephant Bird of Madagascar, the Rhea of South America, the Cassowary of New Guinea and the Emu of Australia. None of these Ratite groups seem to have developed as many species as did the Moa (11), nor have they grown as tall as the largest Moa, *Dinornis giganteus*, which attained three metres (ten feet) when standing erect and weighed about 200 kilograms (440 pounds). Following the extinction of New Zealand's dinosaurs at the end of the Cretaceous period, Moas were able to evolve into an impressive range of plant-eaters, capable of browsing vegetation at a variety of levels in a wide range of forest, shrubland and tussock-grassland habitats. Because birds do not have teeth for chewing, Moas encouraged plant digestion by swallowing up to five kilograms (11 pounds) of pebbles (usually quartz because of its hardness) as gizzard stones to grind down the tough cellulose and lignin-rich fibres. They would have had only one significant predator before the arrival of human beings, the awesome New Zealand Giant Eagle (*Harpagornis moorei*), now extinct; weighing about 10-13 kilograms (22-29 pounds), this was the world's largest eagle.

There are many questions about the origins of the plants of New Zealand. How many of them are ancestral Gondwanan forms, and how many subsequently rafted here on the West Wind Drift or arrived on the feathers of birds blown here by the westerly winds? Although the total number of species of higher plants (flowering plants, conifers and ferns) is not high, 2,450, overall 84 per cent of them are endemic. For some genera with heavier seeds, or seeds and spores which cannot survive in saltwater, the figure approaches 100 per cent, indicating their more ancient continental origins. In particular, one fossil Mesozoic plant, *Glossopteris*, is widespread throughout the southern continents and Tasmania, and has been found in marine sedimentary rocks in Southland. Its occurrence provides conclusive plant evidence for at least part of the greater New Zealand landmass having been once part of Gondwana.

Many puzzles still have to be solved if we are to know the full story of New Zealand's flora. The majority view, however, is that

New Zealand's rainforests mix Gondwanan species, like the podocarp trees in the background, and later additions from mainly subtropical families, like the Cabbage Trees (*Cordyline australis*) in the foreground.

an ancestral continental (Gondwanan) element has been enriched with many additions from subtropical families. Furthermore, many of the genera, such as *Coprosma*, *Hebe*, *Pittosporum* and *Myrsine*, have 'adaptively radiated' by producing a wide range of species that can fill a diversity of ecological niches, often from sea level to mountain tree-line.

One of the impressive features of the flora is the remarkable assemblage of alpine plants, but it is New Zealand's forests that are most like the ancestral Mesozoic vegetation of Gondwana, and these same forests are still the preferred habitat for many of the lower animals of Gondwanan lineage. Driving through the lowland rainforests of South Westland, or walking through the vaulted interiors of Whirinaki or Pureora forests, the visitor can easily conjure up images of what the rainforests of Gondwana must have looked like. Entering these living cathedrals, monuments evoking a bygone era when dinosaurs still roamed the Earth, is a priceless experience. To protect them as part of the world's global heritage is a privilege which New Zealand shares with only a few other fragments of Gondwana – places like south-west Tasmania, the forested highlands of New Guinea and the tattered remnants of New Caledonia's forests, with their rich store of archaic plants.

Three main groups of trees – Kauri, the podocarps and Southern Beech – provide this forest link with Gondwana. A characteristic feature of each group, all of whose species are endemic to New Zealand, is the poor dispersal ability of their seeds, especially across seawater. This factor, taken along with their presence in

Northland's remaining forests are dominated by the Kauri (*Agathis australis*), New Zealand's largest conifer, a tree of awesome size with a lineage dating back to the Jurassic period.

the New Zealand fossil record, strongly indicates their status as survivors on the journey of the 'ark'. The Kauri (*Agathis australis*) has the most ancient lineage, dating back to the Jurassic period; because of its superb stature, the enormous size of its unbranched trunk and the quality of its timber, it is the best-known of the 20 *Agathis* species, most of which occur in the south-western sector of the Pacific Ocean. It is the only New Zealand member of the Araucariaceae family, a group of impressive largely subtropical Southern Hemisphere trees which includes also the Norfolk Island Pine, the Monkey Puzzle Tree from Chile and the Australian Hoop Pine. The Kauri is found only in the warmer parts of the country, such as Northland, Coromandel and Great Barrier Island.

Like the Kauri, the podocarps are conifers, but these members of the Podocarpaceae family have only insignificant seed-bearing cones. The name 'podocarp' means 'seed with a foot', referring to the coloured, fleshy, nutritious stalk on the end of the seed (although in Miro and Matai the seeds are completely enclosed in the 'fruit', as in the angiosperms). This 'foot' is attractive to birds, which ingest the seed as well, thereby ensuring the widespread distribution of the seeds in their droppings. This method of distribution is similar to that which later evolved in the true fruit-bearing flowering plants. The podocarps are primarily a Southern Hemisphere family, a reflection of their dominance in Gondwana during the Cretaceous period. They are well represented in the New Zealand fossil record, and the 17 surviving species include the major forest trees – Rimu, Matai, Totara, Miro and Kahikatea, as well as a group of dwarf trees that form a very distinctive 'bog pine' community in wet, infertile sites. One of the latter, the ankle-high Pygmy Pine (*Lepidothamnus laxifolius*), is considered the smallest conifer in the world. Another interesting podocarp subgroup are the Celery Pines (*Phyllocladus* genus), small trees like Tanekaha and Mountain Toatoa whose 'leaves' are really flattened stems.

The Southern Beeches belong to the genus *Nothofagus*, one of the eight groups making up the Beech family, the Fagaceae. Most members of the Beech family are deciduous trees found in the temperate regions of the Northern Hemisphere, but *Nothofagus* has strong links with Gondwana and is confined to the Southern Hemisphere. The New Zealand beeches are evergreen (non-deciduous), and fossils of their precursors date back to the beginning of the Cretaceous. Their present-day distribution reflects the pattern of former land-bridges between New Guinea, Australia (including Tasmania), New Caledonia, New Zealand and Chile (they are also found as fossils in Antarctica). Species of Southern Beech form extensive tracts of cool-temperate forest in each of these countries, especially in montane areas. In New Zealand there are four *Nothofagus* species, commonly called Red, Silver, Hard and Black Beech; Black Beech also has a variety, Mountain Beech, which is frequent on poorer soils and in cold sites. In places beech forests form an almost single-species canopy above a forest floor that lacks the diverse shrub layer so common in the podocarp and mixed-broadleaf forests of warmer, lower altitudes.

At the other end of the scale from the giant trees of the forest are a number of curious lower plants which, like their counterparts in the animal kingdom, have very primitive features and a Gondwanan ancestry. Some are curiosities, like *Dawsonia superba* (the largest moss in the world), the giant liverworts and the remarkably developed lichens on both trees and rocks. Descendants of the earliest leafless land plants, the lycopods (clubmosses) and Psilotales (horsetails), are plentiful. One of the lycopods, *Lycopodium varium*, can grow to 1.5 metres (nearly five feet) in length, hanging in profusion beneath large tree epiphytes. Among the most ancient of plants is *Psilotum nudum*, often found extending from rocks or the forks of trees, its leafless, many-branched stems looking like pieces of seaweed or green coral thrown up onto the land. But perhaps the most botanically renowned of New Zealand's 'living fossil' plants is the forest epiphyte *Tmesipteris*, a member of a genus restricted to the south-west Pacific. Like *Psilotum* this is a relic of the earliest vascular plants (i.e., plants capable of conducting nutrients and water along their stems). Both *Psilotum* and *Tmesipteris* have a direct link back to the primitive plants that colonized the Earth's landmasses about 400 million years ago.

Profuse epiphytes – like this 'living fossil', *Tmesipteris elongata*, from a fern-ally genus restricted to the south-west Pacific – give the podocarp/broadleaf forests a strong affinity to tropical rainforests.

The Journey of Ancestral New Zealand

Why did ancestral New Zealand break away from the Australia-Antarctica fragment of Gondwana 80 million years ago? What were the major events that shaped this new 'continental archipelago' during that long journey in isolation from the other continents? It is all too easy to take the biological analogy with the biblical Noah's Ark too literally, and visualize the New Zealand ark and its plant and animal cargo sailing serene and unchanging above the turmoil to their eventual destination. The reality was quite different. To understand the nature of the wild land that confronted the first human inhabitants, it is worth examining the main geological evolutionary forces that shaped these 'ancient islands'.

Our current understanding of the journey is based on the concept of plate tectonics. The Earth's continents are huge slabs of lighter continental crust rafting along on plates of mobile (and heavier) crust making up the seafloor of the oceans. These oceanic plates are constantly being added to by lava erupting from the mantle, usually along mid-oceanic ridges, and then spreading laterally in the process known as seafloor spreading. During the ice age of the Permian period (280-225 million years ago), glaciers and rivers dumped vast quantities of erosion products on the sea floor off the eastern coast of the Australian and Antarctic sectors of Gondwana – sediments that would eventually form the backbone of New Zealand and New Caledonia. As the floor of the Pacific Ocean (the Pacific Plate) spread westwards, it was pushed under (subducted under) the eastern margin of the Gondwana continent. But the subducting ocean floor was unable to carry the lighter continental sediments back down into the mantle. Instead, they were piled up along the continental margin of Gondwana in a major phase of mountain-building. By the dawn of the Cretaceous period (about 135 million years ago), these enormous frictional forces had pushed the ancestral New Zealand landmass above the sea for the first time. This must have been a continental margin of great diversity in its landforms and wildlife habitats. Fossils from this period tell of a forested land rich in ferns and coniferous trees.

Yet time is the great leveller: for the next 100 million years these mountains were eroded down to mere stumps, and the sedimentary rocks deeply weathered, gradually producing a layer of leached, impoverished soils over the land. By the beginning of the late Cretaceous period (about 90 million years ago), the original mountainous landmass had been reduced to a roughly level plain. (Remarkably, a remnant of this ancient peneplain still survives in the modern New Zealand landscape – the smooth-surfaced uplands of Central Otago, lying around the headwaters of the Taieri River between Rough Ridge and the Knobby and Lammerlaw ranges.) Thus, even by the time it was 'launched' by the seafloor spreading that began to create the Tasman Sea about 80 million years ago, the 'ark' already had a very subdued profile. By the early Paleocene (60 million years ago) an elongated 'New Zealand', extending up through the Norfolk Ridge to New Caledonia, had reached its current distance of 1,800 kilometres (1,120 miles) from the Australian landmass. Further seafloor spreading in the late Eocene (40 million years ago) finally broke the land-bridge to New Caledonia, the tiny fragment of Gondwana that retains the greatest affinity with New Zealand.

Early in the Oligocene epoch (35 million years ago) the 'New Zealand continent' was a vast shallow sea streaked with a few elongated, low islands. Alpine habitats had disappeared, and the smallness of the islands meant that many land species were lost through competition for the available space. Biological diversity was reduced markedly. It is likely that one of the reasons for the

ABOVE The Nikau (*Rhopalostylis sapida*) is New Zealand's only member of the Palm family. It established itself here during the Miocene epoch, when the climate was distinctly tropical.

BELOW The Yellow-eyed Penguin (*Megadyptes antipodes*) is a species endemic to New Zealand. Penguins evolved in the Southern Hemisphere to become birds adept at 'flying' underwater rather than in the air.

curious geographical distribution of some of New Zealand's lower animals, such as land-snails, may date from this time, with distinct populations becoming isolated on what were then different islands. Not only was the profile of the land eroded down, the land was also severely impoverished of nutrients, the result of eons of climatic weathering and leaching of soils by organic acids dissolved from plant leaves, stems and litter. Ultimately the rocks and soils were reduced to an infertile residue of quartz-rich sand and inert iron and aluminium oxides. It is ironic that this time of acute crisis for so many of the Gondwanan terrestrial plants and animals coincided with a burgeoning of marine life.

At the beginning of the Miocene epoch (24 million years ago) the 'New Zealand' climate was distinctly tropical. The Nikau Palm established itself, and mangroves must have thrived on the margins of the warm, shallow seas. Corals grew along the coasts of what were to become Northland and Auckland. However, this state of affairs was not to last, for progressive seafloor spreading further south was setting up the conditions for the development of a frozen polar continent, Antarctica, which would henceforth dominate the climate of the Southern Hemisphere.

As the gap between the huge continental barriers of Australia, Antarctica and South America widened, the Southern Ocean grew in extent, and with it the circumpolar current – driven west-to-east by the strong winds that developed at these temperate latitudes (winds today known as the 'Roaring Forties' and 'Furious Fifties'). This combined motion of wind and surface ocean current probably accounted for the largest immigration to New Zealand of Australian species – plants like *Celmisia*, *Sophora* (Kowhai), *Epilobium* and the well known Australian Bush trees, Wattle and Bluegum (these latter would prove unable to survive the cold of the subsequent ice age). The Australian element in the birdlife of modern New Zealand – birds like Takahe and Pukeko, Bellbird, Tui and Kaka – probably also arrived around this time, riding the West Wind Drift. New Zealand now extended much further south than Australia, and was probably preferred by penguins, albatrosses and seals because of its cooler habitats. Nevertheless, the West Wind Drift remained the physical link between these new lands of Antarctica, New Zealand, southern continental Australia, Tasmania and southern South America. Even though the continents were drifting apart, their winds continued to originate in the west and the clockwise circum-Antarctic current lapped their shores, constantly cycling organisms and maintaining similar cool-temperate climates.

A sustained period of mountain-building that began about five million years ago and continues today has uplifted the Kaikoura Mountains (shown here), the Southern Alps and the axial ranges of the North Island.

The close of the Miocene epoch not only heralded the onset of the cool-temperate climate that prevails in New Zealand today, it also witnessed the development of a major crack in the Earth's crust – a crack that developed into the Alpine Fault (and associated faults) along the boundary between the great Pacific and Indian-Australian plates. The land again began to buckle and twist, a phenomenon that increased all around the margin of the Pacific about five million years ago (the onset of the Pliocene epoch). Already the Pacific Ring of Fire (and of earthquakes) was emerging – the volcanic arcs of the Aleutian Islands, California and Peru-Chile on the northern and eastern margins of the Pacific, and, along the western margin, Japan, the Philippines, the volcanic island chains of Melanesia and western Polynesia, and New Zealand. The mountainous backbone of North and South America – the Rocky and Andes mountains – began to develop. At the same time, the shape of modern New Zealand began to emerge. Once again, a 'land uplifted high' became a reality, with the formation of a south-west/north-east mountain axis: the Southern Alps, the Kaikoura Ranges and the backbone of the North Island extending from the Rimutaka Range near Wellington to the Raukumara Range near East Cape.

As the land was pushed up to form these mountains, some lowland plants and animals were able to adapt and fill new subalpine and alpine niches. Woody plant genera like *Hebe* and *Coprosma* responded such that the present ecological range of their many species extends from sea level to snow-line. Some of the more cold-tolerant conifers were likewise able to adapt to the high altitudes, producing distinctive subalpine communities of Mountain Cedar (*Libocedrus bidwillii*), Snow Totara (*Podocarpus nivalis*) and Celery Pine (*Phyllocladus aspleniifolius* var. *alpinus*). Elsewhere, alpine plants like the Penwiper Plant (*Notothlaspi rosulatum*) and the Scree Buttercup (*Ranunculus haastii*) developed extensive root systems capable of ensuring their survival on the moving greywacke scree slopes of the mountains of Marlborough and Canterbury. The Rock Wren and Kea (an alpine parrot) probably evolved from lowland parent stocks; the latter's relationship with the other *Nestor* parrot, the forest-dwelling Kaka, is obvious (see photographs, pages 105 and 162). A wide range of specialized insects developed to exploit the new alpine habitats, including the Black Mountain Ringlet Butterfly, which uses its wings like solar panels; day-flying alpine moths; a diverse range of alpine wetas, cicadas, beetles and grasshoppers; and many other insects, including craneflies and moths, which became flightless in order to survive in this cold and windy environment.

LEFT The present terminal face of the Franz Josef Glacier on the West Coast. At the height of the Pleistocene Ice Age, glaciers covered most of the Southern Alps and Fiordland, with valley glaciers extending far beyond the mountain ranges to merge into huge ice-sheets and create the glacial landforms that can today be seen here as well as in the McKenzie Basin and the lake country around Queenstown and Te Anau.

BELOW RIGHT The Southern Alps seen across the mouth of the Waiho River. Since the last ice age New Zealand's glaciers have retreated back into the mountains, but the 'mountains to the sea' glaciation has left its imprint on the landscape: the Waiho flows into the Tasman Sea where the Franz Josef Glacier once did.

The Impact of the Pleistocene Ice Age

The last major chapter in the shaping of the landforms of modern New Zealand occurred during the ice age that began 2.4 million years ago (the beginning of the chilly Pleistocene epoch). Ice ages, and exactly what causes them, are among the great enigmas of the Earth's environment, ranking with the puzzle of why the dinosaurs died out at the end of the Cretaceous. Prior to the Pleistocene ice age the Earth had not experienced a prolonged and severe cooling since the Permian ice age, which ended 235 million years ago. Every region of New Zealand, especially the South Island and the lower North Island, was severely influenced, although probably not as much as the continents of the Northern Hemisphere and Antarctica. During this epoch, conditions alternated between very cold glacial phases (as many as 20) and warmer interglacials. The last glacial phase in New Zealand, generally called the Otiran Glaciation, began 100,000 years ago, peaked between 25,000 and 15,000 years ago, and finished at the end of the Pleistocene – only 10,000 years ago. However, even cold glacial phases like this one had their warmer stages; the four cold stages of the Otiran Glaciation were separated by three warm intervals, each lasting several thousand years, when forest must again have covered much of the land.

At the peak of the Otiran Glaciation so much of the Earth's moisture was tied up in ice in the polar and temperate latitudes that world sea levels were at least 130 metres (425 feet) lower than those of today. Consequently the large harbours of Northland and Auckland were dry land, and many of today's offshore islands, such as Stewart, Kapiti, D'Urville and Great Barrier, were connected to the two main islands. Indeed, the North and South Islands were themselves connected – not across Cook Strait (which is too deep) but via a great land-bridge between Taranaki and the Farewell Spit/Golden Bay locality in north-west Nelson.

This last major glaciation did not produce glaciers in Stewart Island, Central Otago or most of the mountains of north-west Nelson. However, the extent of glaciation along the Southern Alps and in Fiordland was so great that valley glaciers advanced far beyond the mountain ranges. In places these valley glaciers merged into large ice-sheets, such as those that covered the Haast coastal plain, the Te Anau Basin, the Mackenzie Basin and the upper part of the Canterbury Plains. In the North Island, glaciers built up on the volcanic cones of Tongariro, Ruapehu and Taranaki (Egmont), and small valley glaciers filled the heads of the highest valleys in the Tararua and Ruahine ranges. Huge tongues of ice filled the troughs gouged by earlier Pleistocene glacial advances: the fiords and the great southern lakes of the upper Waitaki, Clutha and Waiau catchments. At the same time, glaciers released vast amounts of debris into waterways, thereby building wide 'outwash surfaces' of water-sorted material, or bulldozed the angular blocks of rock into chaotic heaps (moraines) adjacent to the glacier edge.

Two main factors, acting in concert, caused an extraordinary rejuvenation of the surface of the land: one was the abrasive power of the ice, the other the severe reduction of plant cover caused by the cold temperatures (and drier conditions). Glacial ice can split and pluck at bedrock, and some of the resultant fragments are ground down to fine particles called rock flour. This fine, silt-sized material was widely deposited in the valleys below the glaciers by meltwater streams, then swept up by the powerful winds that continually scoured this treeless landscape to be deposited on the surrounding landscape as loess.

The imprint of the ice-age glaciers, therefore, extended far beyond their physical limits – as can be seen today in the smooth contours of the loess-mantled downlands of North and South Canterbury and the uplands of East Otago. One of the outstanding coastal landscapes of the South Island, at Dashing

Rocks near Timaru, consists of 15 metres (50 feet) of cream-coloured loess sitting on top of jet-black polygonal blocks of basaltic lava.

From the air, the courses of these ice-age glaciers can be traced more easily, like the tell-tale silvery trails of a snail on a dewy morning. Forest-covered lateral moraines mark the flanks of many of the great West Coast rivers, like the Whataroa, Waiho, Cook and Karangarua, each now flowing where a glacier once existed. These moraines are now truncated by the sea, which progressively swept back across the land as the ice began to melt at the end of the ice age. The Omoeroa Range in Westland National Park is one such ancient glacial-outwash surface, now covered with the snaking lines of moraines from glaciers long gone. Closer to Franz Josef township, State Highway 6 crosses the Waiho Loop, a younger moraine (about 11,000 years old) whose gently curving arc marks the front (terminal) of one of the last advances of the Franz Josef Glacier at the close of the ice age. A string of small forest-girt lakes – Mapourika, Wahapo, Paringa, Moeraki and many others – marks the troughs of past glaciers on the West Coast. On the eastern side of the Southern Alps the glacial lakes are much larger, with 12 of them in all running from Lake Sumner, near Lewis Pass in North Canterbury, to Lake Poteriteri in Fiordland, on the shores of Foveaux Strait.

The fiords are the most spectacular and best known of the ice-age landscapes but a wealth of others illustrate different interactions between the ice and the surrounding land. For example, in the Matukituki Valley and along the shores of Lake Wanaka the ice has so scored the mountainsides of schist rock that the parallel marks (striations) can still be clearly seen today. In some places, such as at Mount Iron beside the present-day township of Wanaka, or at the junction of the Motatapu and Matukituki rivers, the ice rode up over resistant blocks of schist, leaving isolated mounds called roches moutonnées; the term comes from the French and indicates the resemblance of these distinctive features to reclining mountain sheep. The 'sheep' face down-valley, with the up-valley slope being gentle and smooth and the down-valley face steep and broken where the ice plucked away at it. In other places the retreating glaciers left behind huge blocks of rock (erratics), now incongruously perched among landscapes to which they have little geological relationship. Good examples can be found near Arrowtown (from the glacier that filled the Queenstown Basin) and near The Key, east of Te Anau.

The mountainsides above Lake Pukaki still carry the evidence of the various advances of the Tasman Glacier during the last glaciation. Four different advances, between 60,000 and 12,000 years ago, can be recognized from the four moraine-covered shelves that 'step down' to the present-day shores of the lake. Superb terrace landscapes throughout New Zealand relate to the release of vast amounts of rock debris to waterways after the glaciers began to wane 14,000 years ago; what happened was that, as the sources of the debris became less prolific, the rivers began to cut down into the outwash surface, leaving areas of that surface high (though rarely dry!) as terraces. Flights of such terraces – like the very well developed ones in the Grey-Inangahua depression, the Broken River Basin (near Arthur's Pass National Park) and, in the North Island, the Rangitikei Valley – are outstanding examples of successive down-cutting by rivers during the post-glacial period.

Another graphic consequence of the ice-age climates was the development of extensive tracts of so-called periglacial landscape,

ABOVE Lake Harris, lying in the Humboldt Mountains near the highest point on the Routeburn Track in Mount Aspiring National Park, occupies a hollow gouged by an ice-age glacier – just one of thousands of such alpine lakes.

LEFT On the slopes of the Dunstan Range in Central Otago there still remain solifluction lobes – places where the soil has flowed like treacle and which indicate the intense freeze-thaw environment experienced even amid these ice-free mountains, sheltered as they are in the rain-shadow of the Southern Alps.

ABOVE RIGHT *Gentiana cerina* on the Auckland Islands. New Zealand's alpine flowers are predominantly pale, yet in the equally severe subantarctic environment members of the same genera can be brightly coloured.

particularly in the rain-shadow high country to the east of the Southern Alps – the block mountains of Central Otago (see page 173), the Mackenzie Basin in the upper Waitaki, and the basins of inland Marlborough. These areas were not directly glaciated but were significantly affected by the dry cold. They could no longer support forest, and instead bore at best only a sparse vegetative cover of subalpine herbs, tussock grasses and scattered subalpine shrubs (species of *Hebe*, *Cassinia*, *Coprosma*, *Dracophyllum* and others); in many places the ground was bare. The soils were subject to an extremely demanding daily cycle of freeze (at night) and thaw (during the day). The freezing of meltwater to form ice crystals in the upper layers of soil lifts the topsoil and gradually destroys its structure – a process known as frost-heave. The process goes on inexorably unless there is vegetation or plant litter to shield the soil surface from the cold. Gradually a fascinating landscape develops on the gentler slopes, as the soil and rock fragments are sorted (very slowly, perhaps over several thousand years) into what is called patterned ground, including formations such as soil-stripes (a series of hummocks and furrows), stone polygons and stone drains. On moderate slopes, where drainage is impeded, the soil can 'flow' like ripples of treacle poured from a jar – a process called solifluction. The best examples of periglacial phenomena (and of alpine cushion plants) are found today on the broad summits of block mountains like the Rock and Pillar, Dunstan and Old Man ranges in Central Otago (see page 173).

At the peak of the Otiran Glaciation (25,000-15,000 years ago), most of the vegetation and wildlife of New Zealand was, to a greater or lesser extent depending on the proximity of high mountains, subjected to degrees of this periglacial climate. Half of the South Island was then covered in ice, snow and alpine vegetation. Continuous forest (podocarp and Kauri) existed only north of the Waikato and Bay of Plenty, with isolated remnants of beech forest surviving only in the wetter parts of both islands. Most of the remaining lowlands and hill country were treeless, presenting an open, savannah-like landscape of grasses and wiry, small-leaved shrubs. Winds were probably even stronger than they are now, particularly in the Cook Strait region, which must have been exceedingly bleak and treeless. In these circumstances, the subtropical immigrants of the Miocene – the Coconut Palm and the corals, and even the Wattles and Bluegums – could not survive. All of the Protea family, except two species, Rewarewa and Toru, died out, along with *Casuarina* and some podocarp genera. Other casualties were Southern Beeches of the *Nothofagus brassii* species, today found in the warmer climes of New Guinea and New Caledonia.

However, although the environmental conditions were severe, many, possibly most, plants and animals were able to survive. The wide latitudinal extent of New Zealand and the low sea levels allowed migration between what had previously been isolated islands. The temperatures did not plummet overnight, so those plants and animals that could move gradually did so – in terms not only of latitude but also of altitude – in the search for a more equitable niche. As the climate warmed again during the interglacials, woody shrubs and forests reclaimed much of the former open land. In short, as the cold and warm phases oscillated, the waves of vegetation must have marched back and forth – upslope and downslope, north and south, coastwards and then back to the interior mountain valleys.

Some plant groups, however, were able to exploit the onset of colder conditions. As non-forest habitats vastly expanded – in many places extending right down to sea level – these groups radiated into them. This was one among a variety of factors that combined to bring about New Zealand's spectacular and diverse alpine flora, which includes many impressive members of the Carrot family, particularly the Wild Spaniards (*Aciphylla*) and *Anisotome*, with their long spiky leaves and inflorescences up to 1.5 metres (nearly five feet) high. Also present are large numbers and a great diversity of plants belonging to the Daisy family, including the subalpine and generally shrubby *Olearia* and *Brachyglottis*, the colourful *Cotula* (about 70 species), the remarkable 'vegetable sheep' (mat-forming cushion plants, generally on screes) belonging to the genus *Raoulia* and the endemic genus *Haastia*, as well as more than 60 species of *Celmisia*, the familiar Mountain Daisies found throughout the country's mountain ranges. Other common or obvious genera include *Hebe* (about 100 species), *Ranunculus* (the spectacular alpine buttercups), *Epilobium* (Willowherb family), *Gentiana* (Gentian family) and the striking members of the *Dracophyllum* genus (literally 'dragon-leaved' but more commonly termed pineapple scrub because of the shape of the leaves in some species).

One of the most striking effects of the ice-age glaciations can be seen graphically today in the so-called beech gap in the vegetation of the South Island. For 200 kilometres (125 miles), between the Taramakau River near Greymouth and the Paringa River in South Westland, species of beech are absent from the continuous forest cover of the West Coast, and to a large extent this picture is matched on the eastern side of the Southern Alps (although the eastern valleys between the Rakaia and Hopkins rivers are largely devoid of forests today because of fires). This is the highest portion of the Southern Alps, and coincides with the area where the glaciers were at their greatest extent. The most plausible explanation for the beech gap is that, at the height of the Otiran Glaciation, the ice stripped away all the vegetation from this area. Because beech seed is heavy and unattractive to birds, it cannot spread quickly, being limited to short journeys by wind or, by water, along the margins of rivers; it is thus only now that beech is finally making its slow reinvasion from its strongholds in the north and south. By contrast, podocarps and many broadleaf trees were able to reclaim the area long ago because, with seeds encased in 'fruits' that are attractive to birds, they had a significant dispersal advantage over beech in the revegetation of the bare landscape exposed by glacial retreat.

During the ice age the northern and southern ends of the

Southern Alps – specifically the Tasman Mountains and other ranges of north-west Nelson, and the vast massif of Fiordland – were able to act as refuge areas not only for beech but for other plants and animals. Both these areas are major centres of plant endemism, indicating that there were long periods during which landforms and soil enjoyed stability compared with the situation in the central Southern Alps. Valley glaciers must have occurred in both regions, but sufficient ice-free 'ridge islands' remained within each to allow populations to survive. The existence of such isolated 'ice-island communities' is considered to account for many of the curious distribution patterns exhibited today by Powelliphanta land-snails and cicadas in these mountainous regions. We might also speculate that the separation of the Haast and Fiordland species of Kiwi from other 'brown' Kiwi, like the small population still surviving in the forests near Okarito, was accelerated by their physical separation during the ice-age glaciations.

Post-glacial New Zealand Before the Arrival of Humans

New Zealand began to emerge from the grip of the Pleistocene ice age's last glaciation about 14,000 years ago. Sea levels stabilized at about their current mark 6,000 years ago, so the shape of the coastline and the pattern of outlying and offshore islands has been relatively stable since then – save for the localized impact of volcanic activity, such as the formation of Rangitoto Island, and coastal uplift, usually associated with earthquakes. During the early post-glacial period the vegetation and wildlife that had survived the ice age recolonized the land and waters, continually readjusting to relatively minor climatic changes. Major volcanic eruptions persisted throughout the Taupo Volcanic Zone (see page 84), and localized eruptions affected the vicinity of Mount Taranaki (Egmont) and the Auckland isthmus. In particular, four eruptions during this period spread over 100 cubic kilometres (24 cubic miles) of rhyolitic pumice and ignimbrite over the Volcanic Plateau in the centre of the North Island: the Waiohau and Rotoma eruptions, from the Okataina Volcanic Centre, and the Waimihia and Taupo eruptions, from the Taupo Volcanic Centre.

For all but the last 1,000 years of this post-glacial period, New Zealand's landscapes and wildlife were able to come into equilibrium with a more stable climatic environment in the absence of humans. To understand the profound impact of human settlement on this landscape and its wealth of plants and animals, it is worth summarizing the natural character of this primaeval land as it was just prior to the arrival of the first Polynesian peoples.

The overwhelming impression must have been of dense forest, for over 78 per cent of the land area was forest-covered. The alpine zone, with permanent ice and snow, fell-field and snow-tussock grassland, comprised nearly 14 per cent, while lakes, swamps, open riverbeds, dunelands and the eruption-induced shrublands of the central North Island made up a further four per cent. The final four per cent can be thought of as South Island 'drylands': treeless areas with less than 650 millimetres (26 inches) of rainfall per year, and including the open short-tussock basins of Central Otago and the upper Waitaki, the South Canterbury and North Otago coastal areas, and parts of Marlborough around the Awatere catchment.

LEFT Between the end of the last glaciation, about 10,000 years ago, and the arrival of human beings in New Zealand, volcanic activity within the Taupo Volcanic Zone has probably been the most significant influence on the landscapes of the North Island.

Podocarp/broadleaf forest typically has emergent trees standing out above a tight canopy of lesser trees, as seen here on Otago's Catlins coast, where crimson-flowered Southern Rata stands out.

The Podocarp/Broadleaf Forests

As the Otiran Glaciation began to wane swiftly, forests reclaimed the extensive periglacial grasslands and shrublands with startling speed. Birds aided the rapid distribution southwards (and upslope) of different species of podocarps, forming a wide variety of evergreen forest associations – with other conifers like Kauri, Tanekaha (Celery Pine) and the two New Zealand cedars (Kawaka and Kaikawaka), or with broadleaved trees and hardwoods like Tawa, Rata or beech. The most widespread – and, in terms of its structure and species diversity, also the most complex – of these forest associations, generally termed 'podocarp/broadleaf forest', eventually extended from Northland to Stewart Island. Remnants of podocarp/broadleaf forest are what most North Islanders are thinking of when they talk today of the Bush.

The five largest podocarp trees – Matai, Totara, Kahikatea, Rimu and Miro – were found throughout the forest association, each with subtle habitat preferences. Matai and Totara colonized the drier, stonier and less leached soils in the eastern lowlands of both islands as well as many of the coarser volcanic soils. Kahikatea came to dominate two main landforms: the fertile, silty, free-draining floodplains and low terraces of rivers, and the wet margins of the lowland swamps and bogs (generally referred to by the Maori name *pakihi*) of the West Coast of the South Island. Rimu and Miro tended to thrive in a wide variety of situations but especially in the less exposed, moister sites on the Volcanic Plateau and in the wetter western parts of both islands. Rimu formed magnificent dense stands on the glacial outwash terraces of Central Westland between the Hokitika and Karangarua rivers.

The character of this podocarp/broadleaf forest was determined just as much by its structure as by the composition of

TOP The Autumn Orchid (*Earina autumnalis*), one of the more obvious of the perching epiphytic plants that make the structure of New Zealand's podocarp/broadleaf forest resemble that of tropical rainforests.
ABOVE Only one plant of the woody climber *Tecomanthe speciosa* is known in the wild, on the remote Three Kings Islands. However, it has been propagated and is now popular as a garden ornamental.

its trees. By contrast with the deciduous and coniferous forests of Northern Hemisphere temperate latitudes, it was more stratified, its five layers conferring a strong resemblance to tropical rainforests. The highest (**emergent**) stratum, 30-40 metres (100-130 feet) above ground level, usually consisted of Rimu, Totara or Kahikatea and sometimes Northern Rata and Pukatea. The **canopy** stratum, at 20-25 metres (66-82 feet), included Miro, Matai and a wide range of flowering trees: in the warmer north these included Taraire, Towai and Puriri; south of Auckland, Tawa dominated the lowland canopy and Kamahi that of the mid-altitudes; in the South Island, Kamahi was joined by Southern Rata in the wetter south and west; while in the montane forests Kamahi often shared the canopy with another podocarp, Hall's Totara, and with Kaikawaka. The **subcanopy** trees, 10-15 metres (33-50 feet) tall, tended to be more abundant in situations where light could pass easily through the canopy; the tree ferns Mamaku and Ponga were common, along with Mahoe, Pigeonwood, Toro, Wineberry and several species of *Pittosporum*; in coastal areas Kohekohe, Karaka and Nikau Palm were common components of the subcanopy. The **shrub** stratum, reaching to heights of 3-8 metres (10-26 feet), included many plants whose leaves have an unpleasant smell when crushed or are very peppery in taste – in both instances possibly as a defence against Moa browsing. Several species of *Coprosma* were common along with Horopito, Kawakawa, Pate, Five-finger, Rangiora, Heketara and Ramarama. These shrubs sheltered a diverse **ground** layer containing ferns of many genera, mosses, a few grasses and a delightful range of ground orchids belonging to the *Acianthus* (Gnat Orchids), *Chiloglottis* (Bird Orchids), *Corybas* (Spider Orchids), *Pterostylis* (Greenhooded Orchids) and *Gastrodia* (Potato Orchids) genera.

That the podocarp/broadleaf forest which became so widespread throughout New Zealand prior to the arrival of humans had – and still has – very many of the features usually associated with the tropical rainforests of the world was something noted over 70 years ago by the eminent New Zealand botanist Leonard Cockayne in *New Zealand Plants and Their Story* (1910):

> The general appearance and structure of the New Zealand forest is quite different from that of Europe, temperate Asia, and North America. On the contrary, it bears the unmistakable stamp of a mountain forest in a moist tropical country such as Java, and thus comes into the plant geographical class of rainforest.

This characteristic has been explored in much more detail in recent years by another well known New Zealand botanist, John Dawson, in his book *Forest Vines to Snow Tussocks* (1988). Dawson's conclusions are very similar: '. . . New Zealand conifer broadleaf forest probably comes closer to certain montane tropical rather than lowland tropical rainforests.'

On entering a podocarp/broadleaf forest today it is easy to see the strong resemblance to tropical rainforest in the profusion of vines and epiphytic plants. Climbing ferns and lianas (lianes) drape the branches of the trees, and trunks are entwined with root climbers such as the several climbing Ratas with their attractive flowers and leaf patterns. Another common root climber, Kiekie, which belongs to the primarily tropical Pandanus family, is so profuse in wetter forests that its drooping foliage can completely obscure the host tree, which it may have climbed more than 30 metres (100 feet) in search of sunlight.

The most remarkable residents of this very diverse community are the plants that specialize in living far above the forest floor, trading off the risk of drought (or of a gust of wind sending them plunging from their precarious perches) for the certainty of plentiful sunlight nearer the forest canopy. These are the epiphytes, the uninvited 'passengers' which have somehow to manufacture, or scavenge, their own organic soils to survive perched in the boles of trees. Some are ferns, others small shrubs. Species of two genera in the Lily family, *Collospermum* and *Astelia*, are the most spectacular, forming huge 'nests' which perch high in the canopy. But perhaps the most attractive are the epiphytes that hang from these 'nests': species of *Asplenium* fern, *Lycopodium varium*, species of the 'living fossil' *Tmesipteris*, and masses of two of New Zealand's most obvious forest orchids, *Dendrobium cunninghamii* and the Autumn Orchid *Earina autumnalis*, with its unmistakable sickly-sweet perfume.

Kauri Forest

Compared with the podocarps, Kauri was slower to exert its influence as the post-glacial warming allowed forests once more to clothe the landscape. Kauri could not colonize south of a line between Kawhia and Tauranga harbours, because the temperatures were not high enough, but it became locally dominant in Northland, the Coromandel Peninsula and Great Barrier Island. There it formed an intimate mixture with podocarp/broadleaf forest, with purer Kauri forest tending to be concentrated on the poorer soils of the ridges, on old sand dunes or on very weathered clay-rich soils formed on greywacke. Over

Silver Beech forest near the Divide on the Milford Road, typical of the forests of Fiordland. Beech forests dominate the wet mountains of the South Island and the axial ranges of the North Island.

thousands of years, Kauri degrades the soils that it colonizes, because powerful chemicals leach from its leaves, bark and litter to interact with iron and aluminium minerals present in the soil. The ultimate consequence of this process (podzolization) is that most nutrients are removed from the soil; the classic end-result in the sand-dune country is the Kauri 'egg-cup podzol', where a huge 'cup' of white sand – just inert silica sand left as a residue – has formed beneath each Kauri tree. In clay-rich soils the iron and aluminium were often deposited deeper in the soil as an 'ironpan', inhibiting drainage and limiting aeration of the soil – all factors which led to further deterioration of the fertility and structure of the soil.

Such was the stature (and chemistry) of the Kauri that few other trees were able to compete with it. The huge grey trunks were free of branches beneath the overarching crown 35-40 metres (115-130 feet) above the ground. Most of the more northerly species of podocarp/broadleaf forest present – Rimu, Taraire and Northern Rata – tended to be dwarfed and spindly. A few subcanopy trees and ground plants did thrive: they included Neinei (one of the hardy *Dracophyllum* genus, whose species are typical of infertile, acidic soils in *pakihis* and subalpine environments elsewhere), Toru and large tussocks of Kauri Grass (*Astelia trinervia*).

Beech Forest

During the early post-glacial warming the climate seems to have been sufficiently warm to give podocarps and broadleaf trees a competitive advantage over beech. Major beech refuges still existed in Fiordland, North Westland, the Tasman Mountains and the Tararua and Kaimanawa ranges of the North Island. When the climate began to cool again, about 7,000 years ago, the drier, frostier conditions once again suited the expansion of beech forest. For the next 5,000 years beech moved out of its refuges, steadily invading the podocarp/broadleaf forests of the wetter hill country and the Mountain Totara/Mountain Toatoa (*Phyllocladus*) woodlands of the less moist intermontane basins along the eastern side of the Southern Alps.

Why was beech so successful? While the podocarps were able quickly to establish themselves in the grasslands and shrublands with the assistance of birds, beech had to show much more tenacity, and had to gain toeholds within an already forested landscape. However, beech does have the ability to capitalize on local small-scale catastrophes, like landslips, floods and the windthrow of forest trees in storms. Such events expose mineral soils, so that any beech seeds present can germinate in profusion. The other weakness in the podocarp/broadleaf flanks was the waterways. A ribbon of beech was often capable of spreading downstream from a source high in the catchment, and then infiltrating sideways, gradually mingling with the Rimu, Rata and Kamahi of the existing forest. Superb examples of this stealthy colonization can still be seen from the air in the Dean, Rowallan and Waitutu forests of western Southland. Here the Rimu has been reduced to besieged phalanxes of old bronze retainers, strung out along the crests of ridges, as the beech continues to fight its way up out of the valley floors.

Differences within the five taxa of beech enable them to exploit subtle differences in soil depth, fertility and drainage. Red Beech prefers deeper, more fertile soils, especially in the heads of gullies; consequently it has the greatest stature of the beeches. Black Beech and Mountain Beech are both capable of clinging to thin, infertile soils; they tend to predominate in hill country and the mountains, Black Beech on drier sites and Mountain Beech on colder ones. Mountain Beech, the smallest of the beech trees, is often reduced to a small prostrate shrub on wetter, poorer soils at high altitudes. Along the eastern foothills of the Southern Alps it forms an extensive monoculture – the simplest natural forest in New Zealand, and often consisting of same-aged stands, possibly the result of some blanket catastrophe like fire or windthrow. Silver Beech is the most widespread of the species, generally preferring the wetter sites on the western slopes of the axial ranges from East Cape to Fiordland. The gnarled shapes of Silver Beeches, festooned with lichens and trailing moss, provide the quintessential image of the New Zealand forest near the tree-line. Mist seems perpetually to shroud the forest at this altitude – hence its other name, cloud forest. Superb examples of cloud forest can be found above Lake Waikaremoana in Urewera National Park, in the Tararua Ranges near Wellington, or above the western fiords in Fiordland. In Fiordland, Silver Beech shares forest dominance with Mountain Beech, but Silver Beech is so adaptable that it can extend from sea level to tree-line – a tenacity that few other trees can match. Hard Beech is rather the odd species out in the beech community, being the only one that seems to prefer lower altitudes and also the most northerly and discontinuous of the beeches, extending to the Coromandel Peninsula and Northland, although its strongholds are in Wellington, Nelson and North Westland. One very curious isolated Hard Beech community, however, has been found far to the south, on Mount McLean, a granite dome sitting up on the Haast coastal plain near Jackson Bay. These few stragglers probably survived the last glaciation on this tiny granite 'island' by clinging to the upper slopes while the ice of the Arawhata Glacier stripped away all the surrounding forests of the lowlands. Interestingly, the Hard Beeches of this little

ABOVE The Wrybill (*Anarhynchus frontalis*) is unique among birds in that its bill is curved to the right. This curved bill helps it forage under greywacke stones in the riverbeds on the eastern side of the Southern Alps.

BELOW The Tasman River, a typical braided river of glacial origin, flows through the open grasslands of the South Island high country. This is the breeding habitat of the Wrybill.

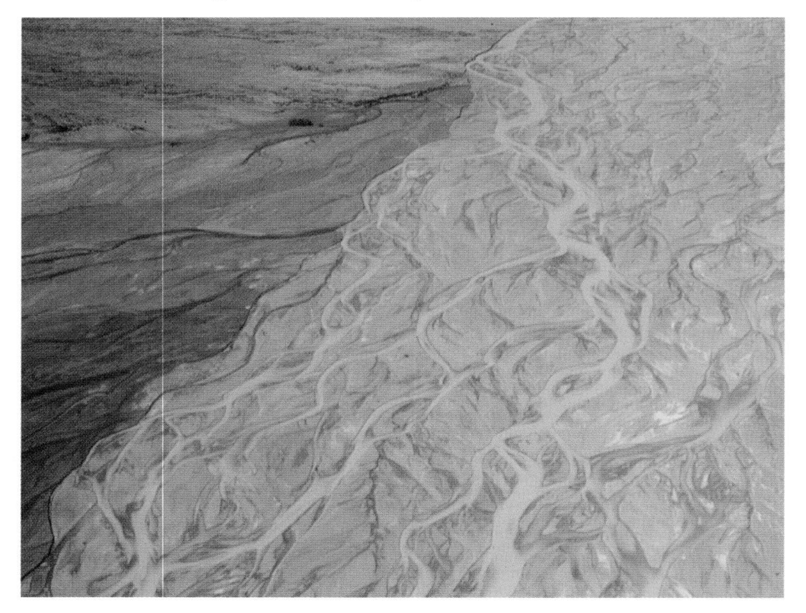

community have not subsequently managed to invade the swamp podocarp forests of the Haast coastal plain (see page 147).

The contrast between beech forest and podocarp/broadleaf forest is very sharp. Beech forests are generally much simpler in structure and lack the diversity of species. Emergent trees are rare (except in mixed forests), and there are usually a tight canopy and a sharp tree-line. In drier areas beech forest is very open and attractive, with light filtering through the canopy to the litter-covered forest floor; lianas (lianes), subcanopy trees and shrubs are limited. Some ground ferns are very characteristic of beech forest: hard ferns (especially *Blechnum discolor*) and the Prickly Shield Fern (*Polystichum vestitum*) abound; the attractive dark green Prince of Wales Feathers Fern (*Leptopteris superba*) is a very distinctive component of wetter, higher-altitude Silver Beech forests; and the translucent Kidney Fern (*Trichomanes reniforme*) forms a carpet, its threads seemingly capable of catching every ray of filtered sunlight. Epiphytes are rare but two parasitic mistletoes – the scarlet-flowered *Elytranthe tetrapetala* and the yellow-flowered *E. flavida* – are a feature of the drier beech forests along the eastern foothills of the Southern Alps. Overall, then, New Zealand's beech forests do not warrant the epithet of rainforest, for they have very few similarities to tropical rainforests. Rather they are the Southern Hemisphere equivalents of the Northern Hemisphere's temperate deciduous forests and cold-temperate coniferous forests.

The Treeless 'Drylands' and Braided Riverbeds

Although the climate quickly warmed during the early post-glacial period, forest was still unable to establish itself in the driest parts of the South Island, generally the more 'continental' intermontane basins where annual rainfall was as low as 350-500 millimetres (14-20 inches) per annum. Moreover, there is plenty of evidence that between 2,500 and 1,500 years ago severe natural fires swept over a wide area through the surrounding dryland forest (mainly Matai/Totara with Kanuka). As a consequence, grasslands and shrublands were induced in most of those areas that had under 650 millimetres (26 inches) of annual rainfall. Likewise, even though the rainfall there was well above 650 millimetres, post-glacial eruptions in the Taupo Volcanic Zone, notably the Taupo eruption of AD186, gave rise to grassland and shrublands. Thus, at various times, much of the central Volcanic Plateau was covered in Red Tussock, Toetoe, Tutu, Manuka and Monoao. In the absence of further eruptions, this vegetation gradually reverted to podocarp/broadleaf forest or, on some sites, beech forest.

Most of the true drylands of the Marlborough, Canterbury and Otago basins were drought-resistant grasslands of 'short tussocks', up to 50 centimetres (20 inches) tall: mainly Hard Tussock (*Festuca novae-zelandiae*) along with Silver Tussock (*Poa caespitosa*) and Blue Tussock (*P. colensoi*). Two rather prickly plants stood out in this landscape; both expanded their range after human settlement, and earned themselves the names Wild Irishman and Wild Spaniard from the predominantly British explorers and pastoralists! The former plant, known to the Maori as *matagouri*, formed tangled groves on the moister slopes of river terraces; its sharp spines are several centimetres long and during the summer, when its branchlets lack leaves, the plant looks as if it were dead. The second 'wild man of the drylands', the Wild Spaniard or *Aciphylla*, was a genus more characteristic of the alpine tussock grasslands, from where it probably migrated as the dryland zone expanded. In the drier areas two leafless brooms, *Carmichaelia petriei* and *C. robusta*, occurred among the short tussocks. On the free-draining alluvial soils of the mountain basins of inland Marlborough, such as the Clarence and Awatere valleys, one of the most beautiful of New Zealand's flowering plants thrived in the harsh unforested landscape: *Chordospartium stevensonii*, the Weeping Tree Broom, which carries profuse pendulous clusters of lavender-coloured flowers from December to February.

The braided riverbeds were small in overall area but formed a very distinctive part of the eastern rain-shadow grasslands of the South Island high country. The ever-changing channels of glacial meltwater could carry many thousands of tonnes of greywacke debris, and developed a wide and intricate network of water and levees. This is a landform unique to the high country and shared with only a few other mountainous regions of the world – the Himalayan and Tibetan plateaux, Alaska and parts of the Andes. It is a harsh and extremely vulnerable habitat, yet it was colonized by a number of specialized wading birds, such as the Black Stilt (*Himanotopus novaezealandiae*) and the Wrybill (*Anarhynchus frontalis*). A few interesting birds can be found in this type of habitat overseas (including the striking Himalayan Ibisbill of the Tibetan Plateau), but the Wrybill is unique among all birds in that its bill is curved to the right. It breeds among the greywacke stones of

The Eglinton Valley. The mountains of Fiordland form an effective barrier to the moisture-laden westerly winds. Eastern valleys in their lee, like the Eglinton, are drier and more open, with Red Beech forest and grassy river flats.

the riverbeds (particularly in the upper Waitaki), where its grey plumage affords it superb camouflage from predator Harrier Hawks and New Zealand Falcons. The curved bill allows it easily to sweep the larvae of aquatic invertebrates from under the rounded stones of the riverbed. After nesting the Wrybills migrate 1,200 kilometres (750 miles) to the estuaries of the Firth of Thames and Northland, where they use their bills to scoop up small crustaceans and worms in the mudflats – a remarkable dual, seasonal function for a bill.

The Alpine Shrublands, Grasslands, Herb-fields and Fell-fields

Some characteristics of the diverse flora of this extensive alpine zone have already been discussed in relation to the impact of Pliocene mountain-building and the Pleistocene ice age on New Zealand's vegetation (see pages 24-9). Today about 600 higher plants (25 per cent of New Zealand's total) are found above the tree-line, and perhaps as many as 500 are restricted to the alpine zone. Furthermore, 93 per cent of them are endemic (compared with 80 per cent for the rest of the higher plants). So, although the alpine and periglacial environment is today relatively inaccessible and much diminished from that existing 14,000 years ago, it is still of as much botanical interest as the forests. The sheer diversity of species and plant-forms – from large snow tussocks and herbs like the Aciphyllas to cushion mats and specialized scree plants – poses a major question for plant biogeographers. How did New Zealand, reduced to a small, worn-down subtropical archipelago during the Oligocene, manage to develop such a remarkable alpine flora in the few million years since the onset of the Pliocene mountain-building? There is no agreement among experts about the solution to this enigma: some consider that refugees from a distant alpine era may have survived throughout the Tertiary; others invoke a land-bridge to Antarctica; and still others point to the high level of hybridism within the alpine flora as evidence of an explosive evolution of new species – many of them from existing lowland genera. Whatever the reason, it is clear that, with the onset of post-glacial warming, the plants of the alpine environment must have engaged in a monumental ecological adjustment – not only with each other, as they strived to exploit new habitats revealed by the melting ice or accumulating loess, but also with the forests and shrublands that had previously been pushed into the coastal and northern regions of the country.

By 2,000 years ago the forests had expanded to their maximum extent and the alpine zone, covering 14 per cent of the country's surface, was much as it is today. It is a gross simplification to equate the extent of this alpine region with merely that of the Southern Alps. They and their northern extension into the Spenser Mountains and Richmond Range became the core of the zone in the South Island, but many other mountainous areas had their own distinctive alpine character. The steep gneiss and granite mountains of Fiordland (including the Takitimu, Livingstone, Longwood and Stewart Island outliers) were relatively low – 1,500-2,500 metres (4,920-8,200 feet) – but were very wet and, around the Darran Range, still glaciated. In sharp contrast, the mountains of the interior 'basin-and-range country' of northern Southland and Central Otago – the Garvie, Eyre, Pisa, Dunstan, Old Man, Rock and Pillar, Kakanui and St Bathans mountains – all lay in the rain-shadow of Fiordland and the Southern Alps. Most were block mountains with undulating summits, and most consisted of schist rocks, which produced a distinctive alpine habitat of slabby rock pavements and rock tors; their climate was more 'continental', and the environment distinctly periglacial in winter (see pages 25-7). Further north lay the frontal greywacke ranges of Canterbury – the Hawkdun, Kirkliston, Grampian, Two Thumb, Ben Ohau, Craigieburn, Torlesse and Puketeraki mountains. They, too, lay in the rain-shadow of the Alps but, with their deep V-shaped valleys and their slopes streaming with screes of greywacke detritus, were certainly anything but block mountains. North-east of here lay the Inland Kaikoura and Seaward Kaikoura mountains of Marlborough; at up to 2,900 metres (9,500 feet) these were New Zealand's highest mountains outside the Southern Alps, and were geologically complex, rent by fault-lines and very dry in summer, although scourged in winter by cold southerly storms because of their proximity to the South Pacific Ocean coast.

On the other side of the Southern Alps, in North Westland and North-west Nelson, a small number of hard-rock mountain ranges – Paparoa, Victoria, Brunner, Arthur and Tasman – were in many ways a mirror image of Fiordland, far to the south-west. As in Fiordland, these mountains were also sharp in outline because of their gneiss, granite and marble rocks. In particular, two striking mountain ranges of red rock, each largely devoid of vegetation, stood out from the rest of the mountains of the South Island: the Dun Mountain/Wairau Red Hills between Nelson and Marlborough, and, far to the south-west, Red Mountain and the Red Hills Range. Their rocks, such as peridotite and serpentinite, are very high in magnesium and iron (hence their common description as ultramafic rocks), and their soils are toxic to many plants. These two curious landmarks emanated from the same geological source and provide an intriguing illustration of the enormous forces that have shaped the mountains of the South Island. For these rocks of Nelson and Fiordland stand on opposite sides of the Alpine Fault, the fissure marking the boundary of the two great plates that New Zealand straddles. So, mountains have been formed not only by uplift on the Alpine fault but also by lateral displacement along it – in this case by 450 kilometres (280 miles) in, perhaps, nine million years.

The post-glacial alpine zone in the North Island was very limited by comparison. The most continuous tract consisted of the axial greywacke ranges of the southern half of the island – the Tararua, Ruahine, Kaimanawa and Kaweka ranges. The two volcanic mountain massifs – Ruapehu-Tongariro and Taranaki (Egmont) – provided localized but higher alpine environments and even, on Mount Ruapehu, small glaciers. The only alpine environment north of latitude 38 degrees South was that on the summit of Mount Hikurangi (1,752 metres [5,748 feet]) in the Raukumara Range near East Cape; this is also the most easterly of New Zealand's mountains.

As forests spread back across the land, beech (with montane podocarps and cedar in beech-gap localities) gradually reclaimed the mid-slopes of the mountains. The alpine communities were

ABOVE The Large Grass Tree (*Dracophyllum traversii*) is among the most distinctive members of its diverse genus, which is widespread in the subalpine shrublands above the tree-line.

BELOW A flowering *Chionochloa teretifolia* snow tussock, one of many alpine plants found only in parts of Fiordland. Snow tussocks are a striking feature of the low-alpine vegetation of New Zealand's mountains.

pushed upslope above a tree-line established at an altitude where mean air temperatures of about 10 degrees Celsius (50 degrees Fahrenheit) were attained in the warmest summer month. This tree-line eventually settled at an upper limit (for Mountain Beech) of 1,500 metres (4,920 feet) for the inland ranges of the North Island (e.g., Mount Ruapehu and the Kaimanawa Ranges) and northern South Island (e.g., the Richmond Range and Tasman Mountains). In the coastal south of the South Island, such as in the Longwood Range or southern Fiordland, it was at 900 metres (2,950 feet) or lower.

On most of the mountains where beech formed the tree-line there was a sharp transition to alpine snow-tussock herb-fields. However, in the beech-free localities – and also at some sites within beech forests, as at valley heads – a tangled mass of mountain shrublands developed instead. In places this almost impenetrable wall, 3-7 metres (10-23 feet) high, of high shrubby species of *Olearia*, *Dracophyllum*, *Hoheria* and *Pseudopanax* extended for 300 metres (985 feet) upslope. It covered the entire southern Ruahine Range and was also well developed on Mount Taranaki (Egmont) and in the western Tararua Range, central Westland and Stewart Island. The three pineapple scrubs, *Dracophyllum traversii*, *D. menziesii*, and *D. fiordense* were striking and colourful features of this shrubland throughout the western side of the main divide, from the Tasman Mountains to Fiordland. Another notable component was the tough-leaved *Olearia colensoi*, its reputation for impenetrability leading to its common name of 'leatherwood', a name first used by the Europeans who explored the Taranaki Ranges and the southern North Island.

The snow-tussock herb-fields above the tree-line (and/or the mountain shrublands) are among the most distinctive elements of New Zealand's alpine landscape, because of the sheer size – 1-2 metres (40-80 inches) tall – and colour – tawny to copper red – of the snow tussocks. Each tussock is like a forest, with each tiller of the tussock an individual 'tree'. They are remarkably long-living

Hugging the ground or wedged into crevices, its leaves felted with fine hair, the South Island Edelweiss (*Leucogenes grandiceps*), found in the rocky fell-fields of the Southern Alps, is superbly adapted to the cold.

Ourisia is one of the many predominantly alpine genera in the New Zealand flora. The flowering herb-fields of the mountains are a major attraction for visitors during early spring.

perennial grasses, larger specimens being several centuries old. Like beech trees they seed infrequently but profusely (probably as a result of being triggered by a warm summer the previous year). Their long, drooping, grooved leaves are superbly adapted for condensing moisture out of mist and then channelling it down to the root base of the plant. All the snow tussocks belong to the genus *Chionochloa*, which has similarities to another Gondwanan relative, the South American *Cortaderia*, a genus that includes the Pampas Grass of Patagonia and the Toetoe of New Zealand, both of which produce tall, showy inflorescences. Of the 14 species of *Chionochloa* in New Zealand's alpine tussock herb-fields, 12 are of tussock form and the other two, Carpet Grass (*C. australis*) and Snow-patch Grass (*C. oreophila*), form a close turf mat, usually at higher altitudes than the tussocks.

Six of the snow-tussock species are particularly important, covering a wide range of landforms and soil types. *C. flavescens*, *C. pallens* and *C. rubra* are the common components of the wetter alpine grasslands in both main islands, although Red Tussock (*C. rubra*) occurs mostly below the tree-line – generally on damp valley floors. This is the tussock found on the volcanic lands of Tongariro National Park and the Moawhango Basin (see pages 86-8). The other three, *C. rigida*, *C. macra* and *C. crassiuscula*, are not found in the North Island; they occur in most of the snow-tussock grasslands of the Marlborough, Canterbury and Otago mountains. *C. crassiuscula* is the main snow tussock of the highest parts of the wet, western tussock herb-fields, while the other two predominate on the drier eastern slopes of the main divide and the interior ranges of Central Otago.

The herb-field element of this snow-tussock herb-field community developed better on the wetter (generally western) slopes of the mountain ranges. The most conspicuous were the larger herbs – again *Aciphylla* and *Anisotome* – and the *Ranunculus* buttercups, the *Ourisia* and *Euphrasia* eyebrights, the *Celmisia* daisies and the *Gentiana* gentians. In the *Chionochloa rigida* snow-tussock grasslands of the eastern mountains the large herbs were much less able to establish themselves, although the Golden Spaniard (*Aciphylla aurea*) preferred this drier habitat.

The higher parts of the South Island mountains – around 1,500-2,000 metres (4,920-6,560 feet) – were able to retain fell-field vegetation communities because of the extreme conditions: cold air, constant strong winds, wide diurnal temperature swings at ground level (cold at night, hot during the day), frost-heave, seasonal snow cover and the freezing of the shallow soils for several months on end. In extreme cases, such as on screes, there were no soils available, only fragments of rock. In such conditions only the toughest of plants were able to establish a (sparse) cover: prostrate shrubs of *Hebe* and, in the lower-altitude herb-fields, hardier species of the main genera – such as *Ranunculus buchananii* and *Aciphylla dobsonii*. A wide range of small cushion plants managed to get a toehold in the drier, rocky fell-fields: *Hectorella caespitosa* and *Chionohebe thomsonii*, their turf mats dotted with flowers in late Spring; *Pachycladon novae-zelandiae* on exposed sites on the schist mountains of Otago, its strong taproot a defence against frost-heave; and the deep green cushions of *Phyllachne colensoi*, which was able to colonize an extraordinary variety of sites, such as snow hollows, exposed ridges and detritus traps between boulders. One of the most attractive of these fell-field flowers is the South Island Edelweiss, *Leucogenes grandiceps*, with its grey woolly stems and orange flower centres so reminiscent of the famous Edelweiss of the Swiss Alps.

Fell-field cushion-plant communities became well developed on the broad summits of the block mountains of Central Otago, where environmental conditions were (and still are) among the most severe in the entire alpine zone of the South Island (see page 173). Further north, on the extensive greywacke slopes of the drier inland Canterbury and Marlborough ranges, species of *Raoulia* and *Haastia* (both genera of the Daisy family) developed into the most spectacular growth form of the high-alpine plants: cushion mats so large and woolly that they came to be called 'vegetable sheep' (see page 27).

Just how high were the fell-field plants able to range? Three flowering plants vie for this record: *Ranunculus grahamii*, a yellow-flowered buttercup widespread in the Mount Cook region, ranges up to 2,800 metres (9,190 feet), but two straggling sub-shrubs, *Parahebe birleyi* and *Hebe haastii*, have been found growing in rock crevices as high as 2,900 metres (9,515 feet) in the central Southern Alps.

The Bird Fauna Before the Arrival of Humans

Because the fossil record is so poor for birds in New Zealand, and because so many rapidly became extinct after the arrival of humans, it is difficult to be too definitive about the composition of the bird fauna during the post-glacial period. However, it is obvious that the rejuvenation of the soils through uplift and glacial erosion promoted the rapid establishment of shrublands and forests. These in turn must have provided a rich harvest for forest birds – nectar and berries as well as copious insects, snails, spiders, frogs and lizards.

We know that the old travellers on the 'ark' survived the ice age – the Moa, Kiwi, Wrens (six of them, although only two survive today) and the three wattle-birds (Saddleback, Kokako and the now-extinct Huia). A few other familiar birds which arrived late in the Tertiary period – Blue Duck, Takahe, Kaka/Kea, Kakapo and New Zealand Pigeon – also survived and thrived. Another arrival during the Tertiary was the Piopio, related to the New Guinea Birds of Paradise and the Bowerbirds of Queensland, but it has become extinct – probably within the last 25 years.

It is likely that most of the seabirds were already well established as the post-glacial shoreline stabilized, for New Zealand's seabirds generally have a longer history than the land-birds. Among them penguins would have been of pre-eminent antiquity, probably having evolved from petrel stock in the late Cretaceous. The high level of marine productivity in the Southern Ocean continued unabated, allowing ample opportunity for seabird populations to expand and diversify. In the words of the eminent New Zealand naturalist, Sir Charles Fleming (from *George Edward Lodge: The Unpublished New Zealand Bird Paintings* [1982]):

> . . . the New Zealand subregion, which extends from the Kermadec Islands to Macquarie Island, became the home of the greatest concentration of breeding seabird species in the Southern Hemisphere. Ten of the world's eighteen penguin species, twelve of the seventeen kinds (species and subspecies) of albatross, forty-four of the hundred or so kinds of petrel, and fourteen of the fifty or so kinds of shag breed in the New Zealand archipelago. They colonized at different times and from several different sources – some from Asia via Australia, others down the West Wind Drift.

Others that we know as native New Zealand birds today – honey-eaters like the Tui, Bellbird and Stitchbird, or the New Zealand Robin – probably arrived at some time during the late Pliocene or early Pleistocene. But a whole host of birds have become extinct during the post-glacial period, virtually all of them through human interference in the form of habitat destruction, hunting or the introduction of predators, competitors and disease.

Over the last few decades, exciting discoveries have been made of sub-fossil remains (usually bones in cave systems) of some of these extinct birds. Among these the Adze-bill and Giant Eagle, *Harpagornis*, have already been discussed (see page 19); others included a large flightless goose (*Cnemiornis*), rails, pelicans, a swan, a large sea eagle and a number of avian predators including the Laughing Owl and the Owlet-Nightjar. Particularly rich concentrations of Moa bones – up to 2,000 individual birds per hectare (810 per acre) – have been found in some eastern South Island swamps, indicating that these huge birds enmired themselves with extraordinary frequency.

Yet, of all the birds that expanded into the shrublands and forests of post-glacial New Zealand, the Moa must have been the most suited to exploit the different habitats. It is now generally agreed that there were 11 species of Moa, belonging to six genera. They ranged in size from the relatively small *Euryapteryx curtus*, weighing 15-50 kilograms (33-110 pounds), to the enormous *Dinornis giganteus*, up to three metres (ten feet) tall when standing erect and estimated to have weighed up to 270 kilograms (nearly 600 pounds). The scope of their habitats was just as wide: coastal and lowland forests, open grasslands/shrublands and forest margins, montane forests and subalpine shrublands – wet and dry, both North and South Islands. The diversity of Moa species (as for the Kiwis and Wrens), compared with the paucity of varieties among more recent arrivals, reflected their long residence in ancestral New Zealand. This diversity could have been even greater, for we do not know how many further members of these more ancient bird groups vanished without trace as a consequence of the environmental stress of the Oligocene inundations or the Pleistocene ice age.

Because of the long association of the Moa species (all herbivores) with the vegetation of New Zealand, there is speculation as to how much evolutionary impact they must have had on the plants. The presence of fibrous juvenile leaves (in trees like the Lancewoods) and spine-tipped leaves (in the Wild Spaniards) have variously been invoked as adaptations. However, the most interesting (and controversial) theory is that the unusually high number of divaricating shrubs and small trees reflects an evolutionary response to Moa browsing. Divarication is the adoption of a form whereby the tough stems become densely interlacing and the leaves become very small, possibly entirely absent from the outer branches. Such plants are almost unique to

'New Zealand . . . became the home of the greatest concentration of breeding seabird species in the Southern Hemisphere. . . . They colonized at different times and from different sources – some from Asia via Australia, others down the West Wind Drift.' – Sir Charles Fleming.

New Zealand; at least 54 species (from 20 genera) are considered to be divaricating, most belonging to the genera *Coprosma*, *Olearia*, *Pittosporum* and *Melicytus*, all of which also have non-divaricating forms. Whereas spines are a defence against soft-nosed browsers like mammals (not yet present), they would do little to deter the large, powerful beak and relatively small head of a Moa. On the other hand, a Moa would have had to expend considerable effort consuming fibrous stem in order to get enough nutritious leaves for itself from a divaricating shrub. The 'deterrent' theory has a plausible ring.

On the eve of the arrival of human beings, New Zealand must have been a land of diverse bird habitats. There was a long, varied coastline with plentiful marine and estuarine life; forests ranged from the coast to the tree-line, from the Kauri forests of the north to the Rimu/Rata forests of Stewart Island; and there were plentiful wetlands, grasslands and open riverbeds. The situation has been well summarized by three of the leading scholars of prehistoric New Zealand:

> From the evidence we have, the postglacial avifauna of New Zealand was rich. There were large and small ground-dwelling herbivores, insectivores and omnivores. Penguins and other seabirds lived around the coasts, and petrels probably nested inland in vast colonies. With no mammalian predators on the ground, but avian predators in the trees, loss of flight was not a handicap, nor was great size. Birds could feed like cows or horses, and be nearly as big. Those species which had to live in the trees evolved good camouflage, those on the ground were cryptic, nocturnal, or both. The conditions allowed the evolution of a highly specialized avifauna, but one which was extremely vulnerable.
>
> GRAEME STEVENS, MATT MCGLONE AND BEVERLEY MCCULLOCH, *Prehistoric New Zealand* (1988)

The Arrival of the Maori and the Pakeha

Sometime between AD750 and AD900 the first wave of Polynesians landed on the shores of the steep and densely forested land that they came to know as Aotearoa. These navigators arrived on double-hulled voyaging canoes (*pahi*) stocked with the plants and animals that were an essential part of their culture. Their journey must have been carefully planned, but was still a calculated risk: one further island-step in a series that built into a great migration into the remote islands of the Pacific Ocean – a journey begun 5,000 years earlier when their Austronesian forebears began to sail eastwards from the shores of South-East Asia. The exploratory skill mastered by these Neolithic people is breathtaking: to Peter Buck they were the 'Vikings of the Sunrise'; to Elsdon Best they were 'the most daring of Neolithic navigators

BELOW A Moriori dendroglyph (tree-carving) on the Chatham Islands. Traditional Maori culture was rooted in *whanaungatanga*, a holistic relationship between human beings and their surroundings.

and explorers in the history of mankind'. In time they became the Maori of Aotearoa.

What was the 'conservation ethic' of these first Maori? Maori culture is rooted in *whanaungatanga*, a holistic relationship between humans and their surroundings (both living and inanimate). For both sprang from the same source, Ranginui the Sky Father and Papatuanuku the Earth Mother, through the work of a Creator, Io-Matua-Kore. All things natural were the children of Ranginui and Papatuanuku: the ocean and all life within it (Tangaroa), the trees and birds of the forest (Tane-mahuta), the winds and climatic elements (Tawhirimatea), volcanic fire and earthquake (Ruaumoko), the Kumara and cultivated crops (Rongo-matane), and many others. Humans shared this ancestry through the union of Tane-mahuta with Hineahuone, the first woman. He was pre-eminent among his brothers in this story of creation, for he alone had the power to separate his parents, causing Ranginui's tears for his estranged Papatuanuku to fall as rain, continually renewing the living soil and plants that clothed her. Tane-mahuta could thus enable the others to escape the stifling darkness of the embrace of Ranginui and Papatuanuku and so breathe and experience life-giving light.

All natural things – rivers and mountains, trees and birds, insects and precious stones – possessed *mauri*, a universal life-essence. Consequently, any contemplated action which would affect the natural environment could also affect Maori themselves. The *mauri* of both people and nature was recognized and protected through *tapu*, the notion of sacredness. Some forms of *tapu* were temporary, such as *rahui*, the closing of a resource area to hunting or gathering until it had recovered. In a sense, all resource use

ABOVE Traditional Maori carving is New Zealand's outstanding indigenous artform. Most carvings revered ancestors, although one animal, the Tuatara, was held in such awe that it was always depicted faithfully. The wood most commonly used for carving – and for canoe-making – was Totara, its selection from the forest being necessarily accompanied by the removal of *tapu*.

RIGHT A stand of Kanuka (*Kunzea ericoides*) forest and clumps of Toetoe (*Cortaderia toetoe*) near the mouth of the Hokianga Harbour, in Northland. This type of fire-induced vegetation is widespread in the drier eastern lowlands and in coastal areas throughout the country.

RIGHT The Australian Black Swan (*Cygnus atratus*) was introduced in 1864 to control the watercress (itself introduced) choking the Avon River, which flows through Christchurch. As has often happened with introductions to New Zealand, the swans spread, thriving in most estuaries around the South Island and in due course being declared game birds. Today there are more than 70,000 birds on Lake Ellesmere (Te Waihora), near Christchurch.

was carefully regulated because of degrees of *tapu*. The practical conservation effect of this highly ritualized approach to the use of the natural environment meant that the consequences of use (especially the sustainability of the resource) were always carefully considered. Sometimes a family group might be appointed *kaitiaki*, or stewards, of a particular tribal resource or sacred place. Because of this unity with the land with Papatuanuku, Maori called themselves Tangata Whenua ('people of the land'). Through their *whakapapa* the people could trace their ancestry, their relationship to their mountain and river, their canoe, and ultimately their relationship with other living things. Furthermore, the shadows of their *tipuna* (ancestors) were across all the land, and the exploits of those ancestors were commemorated in place-names – names that today convey meaning and character to most of the geographical features of New Zealand.

From a modern scientific perspective it is possible to see where this conservation ethic failed; but this is not to dismiss it, for much of this *tikanga* (knowledge, values and protocols) was remarkably conserving compared with the excesses of the first 150 years of European (Pakeha) settlement and resource exploitation. Rather, the failure indicates the limited technology the Maori had at their disposal to enable them to survive in a very dynamic living environment. The most powerful tool at the disposal of pre-Pakeha Maori was fire, and it was eventually used to devastating effect. For whatever reason – Moa hunting, inducing the growth of bracken (for fern-root food) or clearing paths – during the ensuing centuries fires, deliberate or out of control, swept away the forests of the drier eastern parts of Aotearoa. By the time the Pakeha arrived the forest cover had been reduced from about 78 per cent to about 53 per cent of the land area. These fires particularly devastated the lowland Matai/Totara and Matai/Kahikatea forests that had covered the more fertile soils of modern-day Canterbury, Otago, Marlborough and Hawke's Bay.

As a consequence of the fires and of hunting, the Moa became extinct, probably in about the year 1600. A further 34 species of land-birds became extinct during this period, including the Adze-bill and the flightless goose (*Cnemiornis*). The Giant Eagle *Harpagornis* was another casualty, probably through a combination of the loss of its forest habitats and the disappearance of one of its major prey, the Moa.

The other very significant impact Maori settlement had on the wildlife of Aotearoa was through the introduction of the Pacific Rat, the Kiore. Kiore must have played a significant role in reducing the numbers of a very extensive range of fauna – small birds, bats, lizards, large flightless insects and land-snails – and they probably slashed the number of Tuatara on the mainland. Yet Kiore were accorded a standing in Maori society because of their importance as a food and because they provided a tangible link with the spiritual home of Hawaiki. The *whakapapa* of Maori and Kiore were linked through Tane-mahuta and Hinamoki, another of the offspring of Ranginui and Papatuanuku. Maori and Kiore had journeyed together across Te Moana nui a Kiwa (the Pacific Ocean) for a long time. It is small wonder, then, that attitudes towards creatures like Kiore today exemplify the cultural differences between Pakeha and Maori over exactly how to value a natural resource – either as vermin to be eradicated or as a treasured *taonga*.

In the much shorter period of Pakeha settlement – from 1840 until the present day – the area of native forest has been halved again, to only about 25 per cent of the land. Ten more land-birds have become extinct, including the Huia, the Piopio and the Stephens Island Wren. Compared with the limited number of Polynesian introductions of alien species, the number since 1840 has been immense: over 80 species of mammals, birds and fish and more than 1,600 species of plants. In many places these have completely changed the landscape and ecology.

In the writings of most of the early European colonists, the shrublands, interior volcanic plateaux and tussock grasslands are seen as bleak and uninteresting, and the wild forests as places to be feared. There were some romantics, like the painter and writer Augustus Earle, who in his *Narrative of a Residence in New Zealand* (1832) thrilled to 'the countryside, wild, magnificent, fresh from the land of nature and inspiring thoughts of God', but they were few. Most equated the subjugation of nature, and the establishment of a familiar (English) countryside, with virtue and godliness. In their view, wilderness was without its own values, but was merely awaiting the redeeming power of the axe, fire and the plough.

The loss, on a massive scale, of lowland, wetland and coastal habitats was not the only impact the Pakeha had on the indigenous fauna. The settlers introduced a wide range of pests which soon decimated the wildlife and some vegetation communities. Rats, mice and others were brought unwittingly or unavoidably; but some, like the rabbit and the Brush-tailed Possum, were introduced for food or fur. Many birds and plants, intended to be kept in aviaries and gardens for sentimental reasons, soon escaped to the wild. Other wild animals were introduced for sport – deer, trout and gamebirds being typical examples – but their main impact was on native plants and animals. Many of the native birds were hunted for food, or as specimens for overseas collectors and museums. Coastal and marine mammals, especially the New

Zealand Fur Seal and whales, were hunted almost to extinction even before the main phase of settlement, and both are still recovering to the levels the marine environment sustained prior to the late eighteenth century.

Gradually the wild landscapes of New Zealand retreated to the mountainous hinterland, to the offshore and outlying islands and to the South Island's wet West Coast. Some lowland regions – especially Manawatu, Hawke's Bay, Wairarapa, Waikato, Canterbury, coastal Otago and the Southland Plains – were almost completely transformed to pasturelands of Ryegrass, Cocksfoot and Browntop.

The adaptations that made New Zealand's wildlife so curious, brought about through their lack of contact with mammals – their flightlessness and their large size – had rendered them extremely susceptible to predation. By the 1880s the need for wildlife sanctuaries, and reserves of the dwindling vegetation communities, had become acute.

New Zealand's National Parks and World Heritage Areas

Designating National Parks is a way whereby a nation can 'protect in perpetuity, for all to enjoy' those special places that matter to its culture. National Parks may be natural or historical in their character and values; they are always protected by Act of Parliament and strictly managed in accordance with management plans that attempt to strike the right balance between protection and public use. Parks are an international currency, indicating to an extent the willingness of a people to forgo any extractive use of the resources they contain.

New Zealand has a long and proud history of National Park protection and an international reputation for quality management, including a high level of public involvement and participation by citizens on the parks' policy-making boards. There are 13 National Parks in all, totalling almost three million hectares (11,580 square miles), or 11 per cent of the country's entire area. Four are in the North Island: Tongariro, Urewera, Egmont and Whanganui, with a total area of nearly 400,000 hectares (1,545 square miles). The South Island, with its much higher proportion of wild landscapes, has the other nine: Abel Tasman, Nelson Lakes, Kahurangi, Paparoa, Arthur's Pass, Mount Cook, Westland, Mount Aspiring and Fiordland, with a combined area of 2.6 million hectares (10,000 square miles), about 17 per cent of the South Island's area. The three largest parks – Fiordland with 1,257,000 hectares (4,835 square miles), Kahurangi with 500,000 hectares (1,925 square miles) and Mount Aspiring with 355,000 hectares (1,365 square miles) – lie in the South Island.

The story of New Zealand's first National Park, Tongariro, is a remarkable one. The core of the park, the volcanic cones of Tongariro, Ngauruhoe and Ruapehu, were given to the nation by Te Heu Heu Tukino, chief of the Ngati Tuwharetoa *iwi*, for reasons of *mana whenua*. This was in September 1887, only 15 years after the first of the world's National Parks, Yellowstone in the United States, had been protected by statute. Tongariro has the honour of being the world's fourth National Park and the first to be freely given by an indigenous people. Furthermore, its international significance has been recognized by UNESCO in its dual designation as a World Heritage Area, in 1990 for the natural values associated with its volcanoes and again in 1993 for its value as a cultural landscape (see page 88).

The concept of National Parks and the commercial denial involved did not come easily to a colonial society that was still trying to raise an agricultural Canaan from the ashes of the Bush. Egmont National Park was formed in 1900, but only after a strenuous campaign by far-sighted individuals; even then, the park boundary was drawn along a tight six-mile (9.65-kilometre) radius from the summit of Mount Taranaki and there was no buffer zone around this perimeter: the dairy farms began abruptly where the forest ended, an indication of the black-and-white approach to reserves that New Zealand administrations would hold for the next 80 years.

Tourism to the 'scenic wonderlands' of New Zealand grew markedly during the 1880s and 1890s, and this interest highlighted the need to protect some of these places as parks or reserves. Prime attractions were the Pink and White Terraces (prior to the Tarawera eruption of 1886 – see page 86) and the Rotorua geothermal areas. The Whanganui River was promoted as the 'Rhine of New Zealand' during the 1890s when steamers began to carry tourists, some of them venturing on overland to Tongariro National Park, Taupo and Rotorua. In the South Island the first Hermitage Hotel was opened at Mount Cook in 1884; by 1890 Quintin McKinnon and Donald Sutherland had opened up the Milford Track; and glacier guiding was available at Waiho (Franz Josef Glacier) by the turn of the century. Fiordland was made a National Park in 1904 and Arthur's Pass in 1929.

However, partly as a result of the Depression and World War II, the next several decades saw slow progress in officialdom's designation of parks. Only with the passage of the far-sighted National Parks Act in 1952 was public frustration turned into action with the setting up of the National Parks Authority, park boards and a National Parks section in the Department of Lands and Survey. Several parks were formed in quick succession: Mount Cook in 1953, Urewera in 1954, Nelson Lakes in 1956 and Westland in 1960. When the alpine wilderness around Mount Aspiring was declared New Zealand's tenth National Park in 1964, political leaders stated that this was likely to be the last National Park; and for the next 20 years it seemed this was to be the case. However, four new 'maritime and historic parks' were formed as loose groupings of scenic reserves and historic reserves: Hauraki Gulf Maritime Park, Marlborough Sounds Maritime Park, Bay of Islands Maritime and Historic Park and the Otago Goldfields Park.

Changing government attitudes and the rise of a powerful conservation movement throughout the 1970s led to a major revision of the legislation and the passing of the National Parks Act 1980. Boards lost their executive powers for particular parks but gained region-wide responsibility for the planning of reserves and other protected areas. The mood was for much less emphasis on scenic grandeur, the hallmark of the 'old' parks; instead there was a desire for New Zealand urgently to achieve a protected-area network that would be truly representative of the country's landform and biological diversity. This goal had been superbly expressed in the wording of the 1977 Reserves Act:

> Ensuring as far as possible, the survival of all indigenous species of flora and fauna, both rare and commonplace, in their natural communities and habitats; and the preservation of representative samples of all classes of natural ecosystems and landscape which in the aggregate originally gave New Zealand its own recognizable character.

Two more parks, reflecting this broadening of the National Park concept, were eventually formed: Whanganui (with the strong associations of the river with the Maori people) in 1986 and Paparoa in 1987. Paparoa National Park was a landmark decision, the outcome of a bitter struggle between conservationists and the by then embattled New Zealand Forest Service. It was also the first of the new 'scientific' parks, its boundaries carefully drawn to protect sensitive karst landforms (including some impressive

cave systems) and coastal forests. But in 1987, just as the centennial of New Zealand's National Park system was being celebrated, all park administration was merged into the new Department of Conservation. National Parks, although still very important in the eyes of the public and overseas visitors, were coming to be seen as just one of many ways of protecting the natural and historic heritage of New Zealand. Another park, Kahurangi (incorporating

The crater of Mount Ngauruhoe. With its companion volcanoes Tongariro and Ruapehu, it was gifted to the nation by Horonuku Te Heu Heu Tukino and the Ngati Tuwharetoa *iwi* in 1887, and forms the nucleus of Tongariro National Park, the world's first to be gifted by an indigenous people.

ABOVE The Kahikatea and Rimu forests of South Westland are New Zealand's outstanding relic of Gondwana's Mesozoic swamp forests, and are a key feature of the Te Wahipounamu World Heritage Area.

LEFT Spectacular limestone formations at Dolomite Point, Punakaiki. This coastline and the Paparoa Range's western slopes were protected in 1987 as Paparoa National Park, the first of the modern 'scientific' parks.

virtually all of the former North-west Nelson Forest Park), was eventually announced in 1994, but by then most conservation efforts were directed towards other priorities: endangered species, marine reserves, conservation through resource-management legislation, forging conservation partnerships with *iwi* (Maori tribes) and planning visitor services for the burgeoning numbers of overseas tourists.

The concept of World Heritage – the limited number of natural and cultural sites that could be deemed to be of global significance – did, however, make its mark during the decade 1984-94. Both Fiordland and Mount Cook/Westland National Parks were given World Heritage status as natural areas in 1986. Tongariro National Park, as outlined above, was recognized as World Heritage for both its natural and its cultural properties. Finally, the three World Heritage parks of the south-west South Island were replaced by the much larger Te Wahipounamu World Heritage Area in 1990. As such, they joined with Mount Aspiring National Park and other important conservation lands like Waitutu Forest, the Mavora Lakes and, most of all, South Westland south of Westland National

Park. The nomination to UNESCO of one large area was made possible by government resolution of one of the major forest-conservation battles of the 1980s: the decision to manage for conservation purposes (rather than logging) the podocarp forests of South Westland. At 2.6 million hectares (10,000 square miles), Te Wahipounamu ('The Place of the Greenstone') represents 10 per cent of the area of New Zealand and is one of the world's great wilderness World Heritage sites.

Natural Heritage Conservation in New Zealand

How can one summarize the history of a nation's stumbling towards maturity in the conservation of its natural heritage? So much of the biodiversity (and natural-resource wealth) of New Zealand was squandered through ignorance, greed and short-term expediency, but there is no point in dwelling at length on this irretrievable loss; to a large extent those actions were the products of a society that had different values, and much less knowledge, than today's. However, there were always individuals far-sighted enough to question the conventional exploitative ethic of those days: William Fox, the artist and Prime Minister; Thomas MacKenzie, the explorer and Prime Minister; Leonard Cockayne, the botanist and teacher; A.P. Harper, the mountaineer, explorer and conservationist; Harry Ell, the politician and conservationist; Rua Kenana, the Maori visionary who stressed the values of self-sufficiency and community to the Tuhoe people of Te Urewera; W.H. Guthrie-Smith, the farmer, naturalist and author of *Tutira* (1921); and many, many others. Some, like Richard Henry, the country's first 'conservation officer', worked selflessly and tirelessly for conservation – in Henry's case as caretaker of Resolution Island in Fiordland where, from 1894 until his departure in 1909, he tried (unsuccessfully) to save the wildlife from stoats and other introduced predators. Others were as influential as Prime Minister Julius Vogel, who on introducing to Parliament the Forests Bill of 1874, lamented what was happening to New Zealand:

> New Zealand entirely unsettled – New Zealand in its old wild state – might be very much more valuable, clothed with forest, than New Zealand denuded of forest and covered with public works . . .

The inexorable loss of the indigenous forests was publicly decried, but at the turn of the century the same air of fatalism seemed to hang over the country's natural heritage as over the future of the Maori race: both were doomed to assimilation or extinction, so there was little point doing anything to save them. Not until the formation of the New Zealand Forest Service in 1920, the Royal Forest and Bird Protection Society in 1923 and the Federated Mountain Clubs of New Zealand in 1931 were there any coherent conservation voices in the decisions being made about the use of the resources of the wild lands. And even then the focus was generally on the soft options of reserving more of the uninhabited back country unsuitable for farming and forestry – the twin pillars of the Dominion's development. Small wonder that in 1980 an authoritative DSIR publication, *Land Alone Endures* observed:

> Nationally, a very high proportion of our total park and reserve area is mountainland unsuited to any reasonable 'developed' use at all . . . In real terms this means that less than 0.5 per cent of New Zealand's area has been designated National Park or reserve in preference to a use forgone; there has been very little real sacrifice.

During the 1940s and 1950s, the New Zealand Forest Service was at the centre of a major rural conservaton movement that swept the country – soil and water conservation. Deer populations had increased alarmingly during the 1940s, to the extent that after World War II a Wildlife Service was set up within the Department of Internal Affairs to attempt to control the numbers through hunting. Deer control was eventually passed to the Forest Service, who effectively married this role with that of possum control, revegetation and the wider conservation management, including the fostering of recreation, of the State forests of the steeplands – the so-called 'protection forests'. The first major challenge to the Forest Service's philosophy of multiple use of such areas came in the late 1940s, in connection with the Kauri forests of Waipoua. Public pressure forced the Forest Service grudgingly to establish one of the first in a series of strictly protected areas under an amended Forests Act: the large Waipoua Forest Sanctuary was the forerunner of the forest sanctuaries and ecological areas set aside after scientific surveys following the Beech Utilization scheme and other, subsequent, forest logging controversies that raged throughout the 1970s.

The Forest Service was, however, much more sympathetic to back-country recreation, and during the 1960s and 1970s it established a system of 18 forest parks, mainly in the State forests of the axial mountain ranges of the North Island. The prototype was the Tararua Forest Park, successfully nurtured by the Forest Service as an alternative to the 'Tararua National Park' proposed as a project for New Zealand's 'Centennial' in 1940. Another concept which fitted well into the Forest Service's sophisticated zoning approach to forest management was that of 'wilderness areas', promoted by the Federated Mountain Clubs of New Zealand as a means of retaining the full spectrum of outdoor-recreational opportunity. The term 'wilderness' is one of the most abused in the environmental lexicon, being often shanghaied to imply any degree of wildness. In the New Zealand context, a wilderness area is a large area of undeveloped land of wild character and without huts, tracks, bridges, etc. As such, wilderness areas not only protect wild nature; they also offer wilderness trekkers opportunities for solitude and adventure, discovery and challenge – without any of the trappings of civilization. Two of the major wilderness areas set aside in forest parks were the 40,000 hectares (154 square miles) of Raukumara, in the very rugged Raukumara Range near East Cape, and the 87,000 hectares (335 square miles) of the Tasman Wilderness Area in north-west Nelson.

The first truly national campaign on a conservation issue was the 'Save Manapouri' campaign, begun spontaneously by Southlanders in October 1969 in an eleventh-hour effort to stop the government raising the level of Lake Manapouri in Fiordland National Park. The Electricity Department wanted to raise the level to supply cheap hydroelectricity to the Comalco aluminium smelter at Bluff. The protest spread across the country and contributed to the government losing the 1972 election. Emboldened by the victory (the level of Manapouri was controlled but not raised), a new breed of younger conservation activists spawned a variety of environmental conservation groups, from Ecology Action to the Native Forest Action Council, from small specialist groups of lawyers and scientists like the Environmental Defence Society to umbrella coalitions like ECO – Environment and Conservation Organizations (of New Zealand). Environmental concerns were big political issues throughout the 1970s and early 1980s: nuclear power, 'Think Big' energy developments for Maui gas, the Clyde Dam on the Clutha River, the Aramoana smelter proposal, mining proposals on the Coromandel Peninsula, and many others.

The Arawhata River's Ten Hour Gorge in Mount Aspiring National Park. The upper Arawhata Valley lies in the heart of the Olivine Wilderness Area, a large expanse of untracked mountainous wilderness.

The main conservation campaigns were for legislation to protect wild and scenic rivers, for marine reserves and for the protection of marine mammals and other wildlife from driftnetting and other indiscriminate industrial fishing; most controversial of all was the series of campaigns to save the remaining lowland forests from logging. The key objectives of the forest conservationists were expressed in the Maruia Declaration, presented to Parliament in July 1977; it was the largest petition ever collected in New Zealand (341,160 signatures, from a total eligible population of about two million). In a dozen or more forests, battles raged between conservationists, local sawmilling communities and the Forest Service. Today the beautiful names of these places – Mangatotara, Mamaku, Maruia, Warawara, Waihaha, Whirinaki, Erua, Rangataua, Pureora, Inangahua, Ahaura, Okarito and Waikukupa – read like a conservationist's litany of great forest sanctuaries, some martyred by axe and fire, others saved to be venerated and visited by future generations. Under these trying conditions, a fourth protagonist, the New Zealand Wildlife Service, attempted to carry out surveys of threatened fauna and negotiate wildlife corridors and viable wildlife reserves. Ultimately, government and public patience ran out with the limited vision, conflicting missions and entrenched attitudes of the existing natural-resource agencies. The radical environmental re-organizations of the mid-1980s were set in train.

Conservation in New Zealand Today

The passage of the Conservation Act in 1987, the establishment of the Department of Conservation/*Te Papa Atawhai* and eventually a Conservation Authority (and 17 regional conservation boards) created an entirely new conservation order in New Zealand. The New Zealand Forest Service, Department of Lands and Survey, and New Zealand Wildlife Service were abolished and their conservation functions assimilated into the new department. Their commercial and trading functions were incorporated into a series of State-owned Enterprises. The Department of Conservation was given, in addition, new and important conservation responsibilities: to advocate and raise public awareness of conservation, to promote conservation in district and regional planning (especially coastal planning), to protect marine mammals, to administer marine reserves, and to give effect to the Treaty of Waitangi in its work.

New Zealand's new 'conservation estate', too, is daunting in its extent: nearly 30 per cent of the land area of New Zealand, including all the National Parks, forest parks, outlying island sanctuaries and thousands of scenic, scientific, historic and nature reserves. All native wildlife is managed by the Department of Conservation, as are all the great walks, such as the Routeburn Track and Milford Track. The conservation priorities for the country are determined through a process of public consultation on regional Conservation Management Strategies, one for each of the Department's 14 conservancies. In this it works closely with *iwi*, volunteers and private conservation organizations, particularly the Worldwide Fund for Nature (WWF) and the Royal Forest and Bird Protection Society.

What are the main priority issues in the conservation of New Zealand's natural landscapes and wildlife today? There are probably three, the first two of which are closely related: protecting and enhancing New Zealand's remaining biological diversity; controlling introduced weeds and pests; and managing the rapidly increasing numbers of overseas visitors to conservation lands.

Protecting Biological Diversity

As a signatory to the UN Convention on Biological Diversity, New Zealand is required to develop a national strategy for the protection of biological diversity. On the face of it, New Zealand's historical record on biodiversity has not been good. Since the arrival of human beings

- 44 endemic birds have become extinct
- 90 per cent of wetland habitats have been lost, and
- indigenous forests have been reduced from 78 per cent to 25 per cent of the total land area.

In addition,

- only 10 per cent of the tussock grasslands which existed in 1840 remain today, and
- almost 300 plants and animals are currently threatened.

The response to this situation has been impressive, given the human and financial resources the Government has been prepared to put into natural-heritage conservation during a decade when the New Zealand economy has undergone major recession and restructuring. Recovery programmes are in place for key threatened birds like Kiwi, Kokako and Kakapo, as well as bats, Tuatara, some lizards, native frogs and some large invertebrates (including land-snails). The programmes involve both intensive management of mainland habitats and translocation to suitable island sanctuaries. The success of the Chatham Island Black Robin

Department of Conservation staff at work in forests near Rotorua on a recovery programme aimed at preserving the Kokako. Ship Rats, Harrier Hawks and Possums prey on the eggs and unattended chicks. ABOVE The setting of a mist net from the canopy branches is a skilled operation. TOP RIGHT The captured bird is carefully lifted out of the mist net and banded for identification. ABOVE RIGHT Recording wattle-size, one of the various bodily measurements made. BELOW LEFT To alleviate shock, the birds are often fed sugar-water while being handled. BELOW CENTRE A radio transmitter is attached to the bird to enable its movements to be tracked. The fixture has a weak link so that, should it become entangled to the extent it might harm the bird, the transmitter will pull off. BELOW RIGHT The bird, equipped with its transmitter, is released from the forest floor.

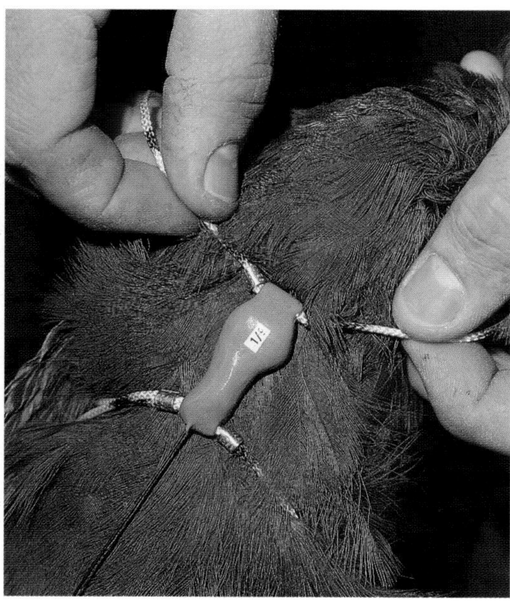

ABOVE The Brush-tailed Possum (*Trichosurus vulpecula*), introduced from Australia, has had a devastating effect on New Zealand's forests, selectively browsing on canopy leaves and all manner of epiphytic and ground-dwelling plants; one native mistletoe has been exterminated.

recovery programme (see pages 198-9) has given managers the confidence to use greater intervention (e.g., supplementary feeding, cross-fostering, captive breeding). However, the emphasis on birds has led to too little recovery work as yet having been done on plants and invertebrates.

The emphasis on saving certain species, albeit at the expense of others, has certainly raised public awareness of the plight of endangered species. But progress has been painfully slow in the establishment of a network of protected areas fully representative of New Zealand's landforms and biological diversity, terrestrial and marine. Although the rate at which marine reserves have been established has been spectacular since the Department of Conservation took over this responsibility, less than one per cent of marine ecosystems are as yet protected. The Protected Natural Areas Programme (PNAP) has simply not received the resources necessary. Two forest-acquisition funds – the Forest Heritage Fund and Nga Whenua Rahui (for Maori-owned forests) – have helped acquire some PNAP priority sites, but semi-arid tussock grasslands, wetlands, estuaries and dunelands are still very under-represented in the protected-area system. Some important tussockland and 'saltpan' communities have been reserved in Central Otago, but satisfactory tussockland reservation in the pastoral high country of the South Island is still a slow business. Resolution will come with careful ecological surveys, followed by negotiation towards freeholding the sustainable pastoral lease-land, and the designation as 'conservation land' of those areas with high conservation value.

A fresh look is being taken at traditional programmes to protect biological diversity. In particular, the relationship between high ecosystem diversity and high landform diversity and/or stability is being considered as a quick way in which key biological communities for conservation may be identified. Areas of high landform stability (and certain rock types) in New Zealand are, in general, much richer in endemic species. Consequently, regions of low tectonic activity (like Northland), or where a high proportion of landforms escaped the ice-sheets of the Pleistocene ice age (north-west Nelson, Marlborough, Central Otago), or which contain particular rock types (e.g., limestone, serpentine), are likely to be priority areas for protection.

The spread of weeds and pests – and their impact on biological diversity – continues to be one of New Zealand's most acute conservation problems. Some predators like rats, cats and stoats are most effectively controlled or eliminated on islands, and this is where much of the work is being done to rescue endangered species. The later sections of this book, dealing with wild places throughout New Zealand, tell further of the spectacular success that scientists and conservation managers have achieved in eliminating tenacious pests from relatively large rugged islands – such as cats from Little Barrier Island's 3,083 hectares (11.9 square miles) and possums from Kapiti Island's almost 2,000 hectares (7.7 square miles). Indeed, New Zealand's expertise and efforts in the field of animal pest control for island rehabilitation are internationally recognized. But on the two main islands rats, cats and stoats, along with browsers like deer and goats, are being controlled only in limited key areas, primarily because of the difficulty and cost of isolating the key area from the surrounding infested land.

Both Common and German Wasps are now major pests in the beech forests of the northern South Island, posing a hazard to wildlife and human visitors alike.The Australian Brush-tailed Possum is present in the New Zealand Bush in epidemic proportions – 70 million animals, found throughout 92 per cent of the country. Not only does their browsing cause immense damage to forest plants but recent research has shown they are also predators upon the eggs and chicks of forest birds such as the Kokako. Despite intensive programmes and increased funding for possum control, it is currently possible to control them on only about eight per cent of the country's conservation lands. Research breakthroughs in biological control are urgently needed if any long-term solution is to be found.

The weed problem is likewise very difficult: Wild Ginger and Old Man's Beard are smothering native plants, pines and *Hieracium* are invading tussock grasslands, willows are choking wetlands and exotic waterweeds are colonizing lakes. There are no simple answers, just the need to press on with urgent applied research, vigilant tactical management, volunteer control programmes and the raising of the awareness of all New Zealanders about what they can do to stop the inexorable destruction of their natural heritage.

Overseas Visitors to New Zealand's Wild Places

New Zealand's wilderness and wildlife – and the 'clean and green' image of its rural landscapes – have brought unprecedented

ABOVE Trampers ascending Mackinnon Pass on the Milford Track – the most famous, just ahead of the Routeburn Track, of New Zealand's 'Great Walks'. New Zealand's National Park system offers a wide range of opportunities for outdoor recreation, from short walks and diving to wild-river rafting and wilderness trekking.

LEFT Conservation-interpreter Derek Brown discusses a Giant Weta with a group of children during a summer-visitor programme on Maud Island. Conservation education is one of the keys to nurturing a generation of people who will not just care about their natural heritage but actively work to protect it.

ABOVE Ecotourists observe Southern Royal Albatross on Campbell Island. New Zealand's management agencies face a dilemma: just how many visitors, how often, can sensitive landscapes and animal communities tolerate?

numbers of overseas tourists to some conservation lands. This is especially so for the Great Walks of the back country and for key scenic areas like Milford Sound, Mount Cook, Cape Reinga and South Westland and the glaciers. A significant proportion of these visitors are nowadays 'ecotourists' (and/or 'geotourists'!) seeking to visit the Bush or an erupting volcano, to see a fossil trilobite or to observe a variety of wildlife: seals at Cape Foulwind, Royal Albatross at Taiaroa Head, whales at Kaikoura or White Heron at Waitangiroto. The New Zealand Tourism Board has set a target of three million overseas visitors by the year 2000 – three times the number who came in 1993.

Conservation lands carry an impressive range of facilities for visitor information, safety and comfort – about 960 huts, 9,000 kilometres (5,600 miles) of tracks, 260 camp-grounds and about 40 visitor centres, most of the latter being located at the entrances to parks, although some are specifically devoted to wildlife, such as the National Wildlife Centre at Mount Bruce in the Wairarapa. Clearly conservation lands and their facilities are a key part of New Zealand's tourism industry. Equally clearly, the tourist industry benefits from the work of the Department of Conservation and its conservation associates, because the sum of their conservation efforts conveys the impression abroad that New Zealand is a place where people care about their environment. Yet it is generally considered that current facilities and services will be inadequate to handle the projected increase in visitors. What have to be worked through are the options for minimizing the physical and biological impacts of such an increase in numbers while at the same time maximizing the enjoyment and the 'learning experience' to be had by each of those individual visitors.

There is a major dilemma here for conservationists and the tourist industry. The economic advantages of tourism were promised as the rewards to many local communities who had to forgo traditional benefits from natural resources, such as logging, mining, fishing and water-exploitation. Ecotourism enterprises operated by local people have sprung up in many of these communities, which indeed are beginning to enjoy the economic blessings. But how much is enough? The growth targets of the Tourism Board will be self-defeating if they result in the degradation of heritage and conservation values.

The best ecotourism in New Zealand, as elsewhere in the world, is likely to be of the 'high quality, low throughput' variety. Most New Zealanders involved in the conservation of their natural heritage are proud of it, and are happy to share their knowledge with interested visitors. But anyone who has introduced visitors to the wonders of the natural environment knows that the quality of the individual visitor's experience can be inversely proportional to the *number* of visitors – and in direct proportion to the amount of *time* spent in the activity.

THE KERMADEC ISLANDS

The Kermadec archipelago consists of a string of volcanic islands lying roughly midway between the Tongan Islands and the north-east coast of the North Island. They are the northernmost land areas of New Zealand, with the largest island, Raoul, lying at about latitude 29 degrees south, some 1,300 kilometres (800 miles) north-east of Auckland. The total area of the islands – under 3,300 hectares (12.7 square miles) – is very small because they are only the tips of deep-sea volcanoes. They lie along the Pacific Ring of Fire, a line of volcanoes that extends on down to White Island and the Taupo Volcanic Zone (see page 84). Raoul and Curtis islands are still active volcanoes, and this whole chain of young islands is one of the most earthquake-prone areas in New Zealand.

The marine environment of the Kermadecs supports a subtropical biota with strong tropical elements, in contrast to the overwhelmingly temperate marine environments of the rest of New Zealand. Although very few of the marine species are endemic, there are exceptions such as the Giant Limpet. The marine life has a number of interesting features: the huge, approachable Spotted Black Groper (Grouper), a lack of tropical herbivorous fish, marine reptiles like sea snakes and turtles, and populations at their southern limits – for example, the Crown of Thorns Starfish and corals (although there are no coral reefs as such). Not surprisingly, this fauna has a strong likeness to that around Australia's Lord Howe Island at similar latitudes in the northern Tasman Sea.

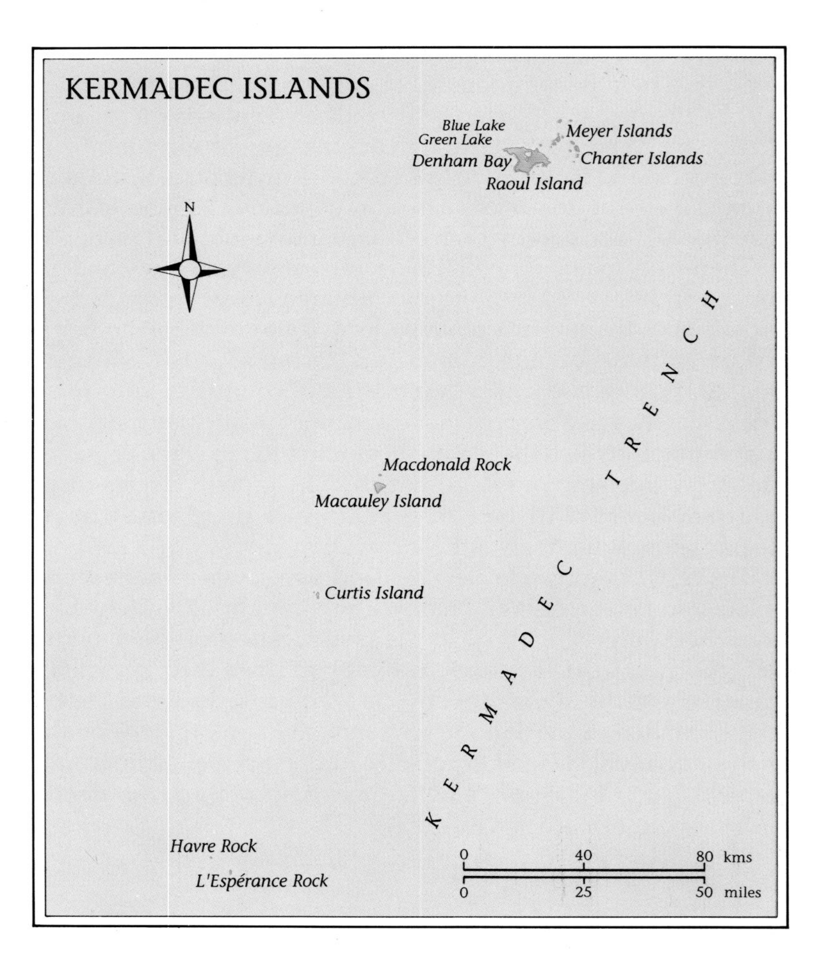

The marine ecosystems of the Kermadecs are probably the most pristine in New Zealand, with no history of exploitation. Their outstanding ecological values were recognized by the designation of a Kermadec Islands Marine Reserve in 1990; at 748,000 hectares (2,888 square miles), it is the second largest protected area in New Zealand (after Fiordland National Park). Marine reserves provide the strictest category of protection for the marine environment under New Zealand's conservation legislation, with fishing and disturbance strictly prohibited within the reserve.

The land flora and fauna of the islands are no less interesting. Like New Zealand's subantarctic islands, the Kermadecs are true 'oceanic islands'; that is, islands that have never been joined to the main 'continental' islands by land-bridges. They have no indigenous reptiles or mammals, and all their plants are assumed to have arrived by wind, rafting across the ocean, or with migrating birds.

A combination of fertile volcanic soils and a moist equable climate has ensured the development of a warm temperate coastal forest on Raoul Island. Most of the forest is dominated by that ubiquitous Pacific genus so at home on volcanic islands – *Metrosideros*, in this instance the Kermadec Pohutukawa (*M. kermadecensis*). Other common forest plants are the Kermadec Mapou, Ngaio, Nikau Palm, Karaka, Wharangi, Kawakawa and some endemic tree ferns. Because of the youthfulness of the islands the total number of indigenous plants is not large (113), but 21 of them are considered to be endemic; some of these, including a hebe and the fern *Cyathea kermadecensis*, are now very rare. The adaptive evolution of distinctive Kermadec species from some of the mainland New Zealand genera that have reached the islands contributes to the international significance of their flora. While there are similarities with Norfolk Island, 1,250 kilometres (775 miles) to the west, much of the flora of Norfolk is now endangered, making the conservation of the natural heritage of the Kermadecs all the more important. Most of the land area of the Kermadec Islands has been protected since 1934; in 1992 the entirety of the islands' land area was strictly protected as a nature reserve.

The Kermadecs were settled by Polynesian migrants perhaps as long as 1,100 years ago; but ultimately this settlement failed. The archaeological relics are considered to be very important for they represent a third adaptation of Polynesian culture to the New Zealand environment (the other two being mainland New Zealand and the Chatham Islands).

Sadly, unlike most of New Zealand's subantarctic islands, the northern Kermadec Islands are now far from pristine. The Maori

Blue Lake, one of two large lakes in the central caldera of Raoul Island. Raoul has erupted three times since the Kermadecs were discovered by Europeans in 1793, the most recent being in November 1964.

ABOVE LEFT Raoul Island is mainly covered with a forest of Kermadec Pohutukawa (*Metrosideros kermadecensis*). *Metrosideros* species colonize lava flows in New Zealand as elsewhere throughout the Pacific.

ABOVE The Chanter Islands and Herald Islets comprise a group of volcanic stacks and small islands lying off the eastern coast of Raoul Island.

LEFT Kermadec Nikau Palm (*Rhopalostylis baueri* var. *cheesemanii*) groves bring a tropical atmosphere to damper gullies throughout Raoul Island.

navigators brought the Pacific Rat (the Kiore), and goats were introduced to Raoul and Macauley islands as early as 1836 to provide food for the occupants of the whaling base set up to exploit the rich whaling grounds in the area. Cats and Norway Rats were eventually introduced as well, through human carelessness. Fires burned much of the original vegetation, and were so extensive on Macauley Island that most of the vegetation was converted to a short cropped turf (with the help of the introduced goats). In addition, since the mid-1800s settlers introduced sheep, cattle and many exotic plants to Raoul.

The introduced mammals multiplied rapidly and, as has occurred on vulnerable islands all around the world, they had an appalling impact on the indigenous birdlife, especially on Raoul Island. The Kermadec Pigeon is now extinct and the Kermadec Parakeet is restricted to the tiny, rodent-free Meyer Islands – two kilometres (1.25 miles) north-east of Raoul – and Macauley Island, where the population appears to be declining rapidly. Likewise, the only population of Spotless Crake within the Kermadecs is on the Meyer Islands. Perhaps the most significant losses were among the seabirds, particularly the huge breeding populations of Kermadec Petrel that once existed on Raoul Island.

The southern Kermadecs, however, are a different story, for

they are generally predator-free. These small islands support very large breeding populations of seabirds, especially shearwaters and petrels. In fact, Macauley and Curtis islands may support the densest seabird populations anywhere in New Zealand. This has led to them being referred to as the subtropical equivalents of the fabulous Snares Islands (see page 182) in the wild cool-temperate climate of the 'Roaring Forties' far to the south of Stewart Island (see page 177).

Today, remote and inaccessible Raoul Island has a small population of Department of Conservation staff (up to ten), some of them short-term volunteers to assist in the weed-eradication programme. The management role for this small, isolated group is very challenging: surveillance of the Kermadec Islands Nature Reserve and Marine Reserve; protecting the archaeological heritage; maintaining a meteorological service; and monitoring the activity of Raoul volcano. Since the eradication of goats from Raoul in 1986 (after a difficult 12-year hunting campaign using dogs) there has been a dramatic change in the vegetation. However, the same benign environmental conditions that are so conducive to the regeneration of the indigenous flora also allow a wide range of subtropical weeds – such as Mysore Thorn, Brazilian Buttercup, Passionfruit and Guavas – to thrive and to pose a serious threat to the restoration of the native vegetation of the islands. Most conservation-management effort in the years ahead will be directed towards eradicating not only these weeds but also the cats and the two species of rats on the islands.

ABOVE Introduced Kiore, cats and Norway Rat have largely destroyed Raoul Island's colonies of Kermadec Petrel (*Pterodroma neglecta neglecta*). This chick is part of a thriving population on the predator-free Meyer Islands, two kilometres (1¼ miles) off Raoul's coast.

BELOW The Kermadecs' pristine marine ecosystems support a diversity of subtropical fish found nowhere else in the country. The waters are protected as New Zealand's largest marine reserve.

NORTHLAND

Northland is very different from the rest of the North Island. It has a warm, humid climate and its forests are dominated by a number of trees, such as Kauri, Taraire, Puriri and Pohutukawa, which are naturally concentrated in this warmer northern region. There are no high mountains; the landscape is generally subdued, with landforms which are older, or more stable, than the rest of the country. The ice ages have left no recognizable imprint and there are no active volcanoes – although the Whangarei and Bay of Islands localities have plenty of small extinct basalt cones. Overall, Northland is an old worn-down landscape, characterized by deeply weathered clay-rich soils, Kauri forests and stretches of sandy coastlines and mangrove-fringed estuaries.

There are a number of interesting wildlife features in Northland. As sea levels have fluctuated throughout the last few million years, there have been times when the entire Northland peninsula has been an island – or several islands, in the Far North locality around Te Paki, Houhora and Karikari. Consequently, endemic species of fauna such as land-snails, stag beetles and perhaps the Northland Green Gecko have evolved in these 'island' localities. The geographic isolation also accounted for the late arrival of some exotic mammal pests, especially possum and deer.

When the first Polynesian voyagers arrived, most of Northland was covered with forest – and 35 per cent of the forest area was dominated by Kauri. Kauri (*Agathis australis*) is one of the great trees of the world. Its height of 45–55 metres (150–180 feet) is sometimes surpassed by Kahikatea and Rimu, but not its girth, which can reach a stupendous 20 metres (65 feet). These impressive dimensions are enhanced by Kauri's ability to shed its lower branches, leaving a massive grey trunk soaring up to 30 metres (100 feet) before spreading into the characteristic upward-reaching limbs of the crown. Equally impressive is the longevity of Kauri, up to 2,000 years, easily the longest-living New Zealand tree. In its great size and age the Kauri ranks alongside the Californian Redwoods, the Tasmanian Huon pines and the Yakusugi (or Japanese Cedar, *Cryptomeria japonica*) of Yaku Island in southern Japan.

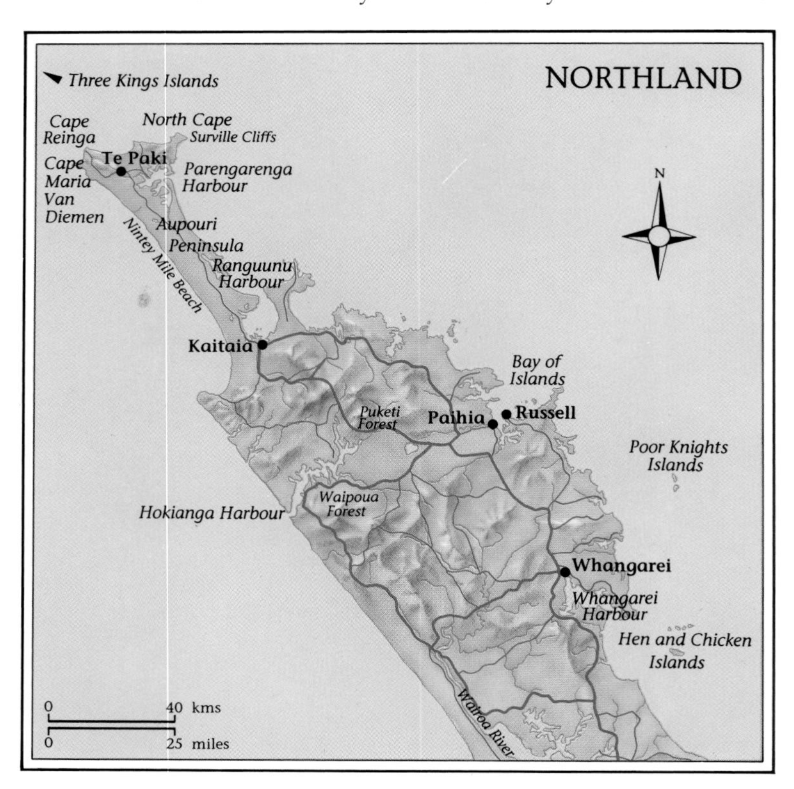

Kauri was so prized for timber that exploitation over the past 200 years has almost extinguished it throughout both the Coromandel and Northland peninsulas. While there were about one million hectares (3,860 square miles) of Kauri forests 1,000 years ago, today there are only 7,500 hectares (29 square miles) of unmodified mature Kauri forests remaining in Northland.

Perhaps even more impressive than the magnificent Kauri trees of Northland are the coastlines. The eastern coast is rocky and deeply indented, with many small offshore islands (especially in the Bay of Islands). Several large shallow estuaries, like Parengarenga, Ranguunu, Hokianga, Kaipara and Whangarei harbours, support teeming populations of wading birds; these include migratory visitors from the Arctic – Godwits, Knots and Turnstones. The estuaries are the stronghold of the New Zealand Mangrove, probably the most distant from the equator of any of the world's mangrove communities.

Two very important island groups, the Three Kings Islands and the Poor Knights Islands, lie off the Northland coast. They are both termed 'outlying islands'; that is, they are at or beyond the limit of the continental shelf, and therefore were not joined to the North Island when sea levels dropped during the last glaciation. Because both groups have been isolated from the mainland for a long time, they have become small centres of plant and animal endemism. This is particularly true for the Three Kings Islands. Both island groups are largely free of introduced mammals and are strictly protected as nature reserves because of their importance as refuges for wildlife and plants not found on the mainland.

On a fine clear day at Cape Reinga, the Three Kings Islands can be seen 50 kilometres (30 miles) to the north-west. Because the islands are influenced by both temperate and subtropical currents, their marine biota is an extraordinary mixture of species that normally frequent either cold or warm waters. A notable feature is the proliferation of large seaweeds (such as the perpetually moving Southern Bull Kelp, *Durvillea antarctica*) to an extent that rivals the waters of the far south of the country. The islands have 13 endemic land plants, including the exceedingly rare woody climber *Tecomanthe speciosa*, of which only one plant is known in the wild; a recovery programme has ensured its vegetative propagation in nurseries and its increasing popularity as a garden ornamental. There is also a high degree of endemism in the invertebrate fauna of the Three Kings, including 25 different types of land-snails.

The Poor Knights Islands are a group of very interesting old volcanic islands lying about 25 kilometres (15 miles) off the Northland coastline. In 1981 the waters around the Poor Knights

Islands were designated New Zealand's second marine reserve, in recognition of their outstanding and vulnerable marine habitats. The extremely high clarity of the deep water, the varied and colourful flora and fauna of the vertical underwater cliff faces, and the many subtropical species of marine life make the reserve one of the most spectacular diving locations in New Zealand.

Northland's rocky eastern coastline is deeply indented by many small coves and offshore islands (particularly in the Bay of Islands). Pohutukawa (*Metrosideros excelsa*) is the most widespread coastal tree on the rocky headlands and cliffs.

The Far North: Te Paki

Beyond New Zealand's northernmost town, Kaitaia, lies the Aupouri Peninsula, 120 kilometres (74 miles) long and connecting the rest of Northland to a group of very old volcanic landforms around Te Paki in the far north. Here, between Cape Maria Van Diemen in the west and North Cape in the east, lie some of the most visually impressive coastal landscapes in the North Island. To the Maori people the coastline around Cape Maria Van Diemen to Cape Reinga is the most sacred of places – Te Rere i nga Wairua – where the spirits of the dead set out for Te Reinga, the Maori spirit-land.

The main attractions for visitors are the golden sands of Ninety Mile Beach (the 'highway' for the tourist buses) and the white sand hills of Te Paki Stream, leading to 'land's end' at Cape Reinga. Further to the south, nestling among a wide expanse of sand dunes, lie the myriad blue waterways of the Parengarenga Harbour, probably the most natural large estuary in New Zealand. In the 40 kilometres (25 miles) of the Te Paki coastline from the northern end of Ninety Mile Beach to the mouth of the harbour are many fine beaches, five of which are outstanding: Spirits Bay, Tom Bowling Bay, Waikuku Beach, Twilight Beach and Te Werahi Beach. Together they make up the most botanically valuable dune systems in the North Island – and possibly in all New Zealand.

The conservation and recreation value of these lands has been recognized by protection within the 25,000 hectares (96 square miles) of the Te Paki Reserves complex. The vegetation is mainly Kanuka-Manuka-Rewarewa shrubland and rushland, an important habitat for the large endemic land-snail *Placostylus*. Of scientific interest is the 'serpentine dwarf shrubland' on the cliffs and plateau between Kerr Point and Surville Cliffs. Because of the chemistry of the soils (which are derived from the local ultramafic rocks) many of the plants are dwarf or prostrate forms of common species like Rewarewa, Tanekaha, Fivefinger and Pohutukawa. This small but remarkable locality probably supports the highest density of endemic plant species and forms in mainland New Zealand.

RIGHT The 2,000 hectares (7.7 square miles) or so of white dunes around Te Paki Stream at the northern end of Ninety Mile Beach are among the far north's most spectacular landforms. They are all that remains of more than 30,000 hectares (116 square miles) of active dunes that covered much of the Aupouri Peninsula until stabilized by progressive plantings of introduced species, especially Marram Grass, lupins and pines. Archaeological investigations of Maori middens show the dunelands once had a significant Maori population.

LEFT Spirits Bay, on Te Paki's northern coast between Cape Maria Van Diemen and North Cape. An eight-kilometre (five-mile) sweep of near-pristine foredunes is backed by the wetlands of Paranoa Swamp and Waitahoro Lagoon; beyond these, the vegetation grades into coastal Kohekohe, Taraire, Hinau and Puriri forest.

ABOVE Pied Shags (*Phalacrocorax varius*) at Spirits Bay. These remarkably tame birds, though primarily marine, may be seen on inland lakes and waterways, perching with wings outstretched on rocky outcrops and dead trees.

RIGHT Cape Reinga – Te Rere i nga Wairua – the focal point of the wild, scenic Te Paki coastline. Here, in traditional Maori belief, the spirits of the dead set out for the spirit-land, Te Reinga. Offshore, the waters of the Tasman Sea and the Pacific Ocean mix in turbulent eddies across the Columbia Bank.

ABOVE The North Cape locality probably has mainland New Zealand's greatest concentration of endemic plants, partly because of the ultramafic rocks and the area's isolation as an island for many millions of years. This shrubland of Kanuka (in flower), Manuka and Ti Kouka (Cabbage Tree) has been burnt regularly since human occupation.

LEFT Several pairs of Pied Shag may nest in the same tree. Their large nests of sticks, seaweed and plant debris are a distinctive sight in coastal Pohutukawa. They nest at any time, depending on food supply, laying up to five eggs.

BELOW *Hebe macrocarpa* var. *brevifolia*, one of many low-growing shrubs found only on the serpentine soils of the Surville Cliffs near North Cape. Many plants in this unique community, with numerous endemic species, have adopted a form whereby the stems do not climb but trail down the cliffs through other plants.

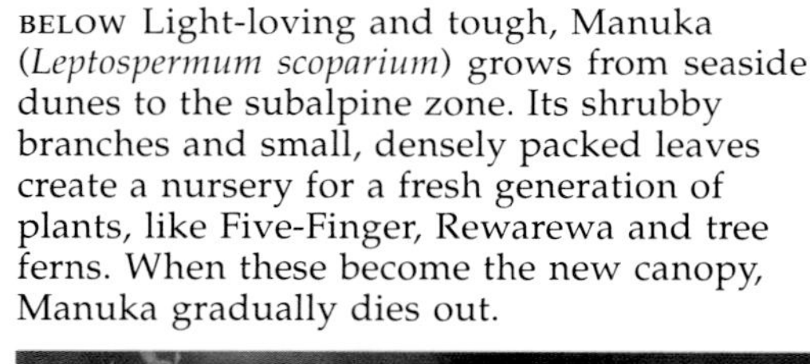

BELOW Light-loving and tough, Manuka (*Leptospermum scoparium*) grows from seaside dunes to the subalpine zone. Its shrubby branches and small, densely packed leaves create a nursery for a fresh generation of plants, like Five-Finger, Rewarewa and tree ferns. When these become the new canopy, Manuka gradually dies out.

BELOW *Placostylus ambagiosus* – one of the three New Zealand species of Flax Snail, all Gondwana relicts and all restricted to Northland – is confined to isolated sites north of Ninety Mile Beach. Flax Snails do not eat flax but prefer foraging in the leaf litter in coastal forest.

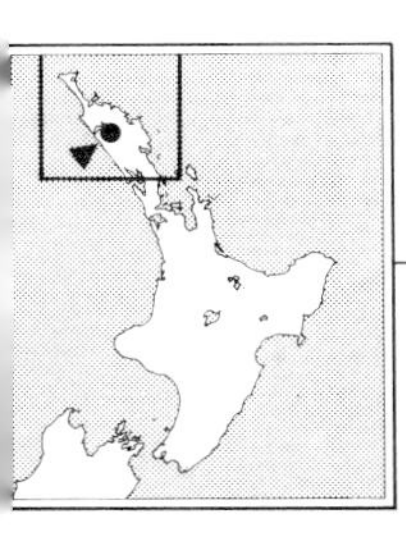

The Great Kauri Forests: Waipoua, Omahuta and Puketi

The best remnants of the once-great Kauri forests of Northland now lie in five upland 'forest islands' – each of area 4-9,000 hectares (15-35 square miles) – encircling the Hokianga Harbour. Waipoua, Waima and Warawara forests lie on the undulating old lava plateaux of the west; Puketi and Omahuta forests on the greywacke hill country at the head of the Hokianga. The Department of Conservation has proposed that these five forests be combined with a further 42 tracts of protected Kauri forest to form a 'Northland Kauri National Park', with an area of over 105,000 hectares (405 square miles). Some of the forests – such as Manginangina, Trounson and part of the Waipoua Forest Sanctuary – are very accessible and contain magnificent Kauri.

The best Kauri forest communities in Waipoua Forest were strictly protected as a 12,800-hectare (49-square-mile) forest sanctuary in 1952 after one of New Zealand's earliest conservation controversies. The drive along State Highway 12 through this forest is an outstanding experience, with huge Kauri, Rimu and Northern Rata towering above a canopy of Towai, Tawa and Pukatea. Recreational highlights include three short walks to magnificent Kauri trees: Tane Mahuta, Te Matua Ngahere and the Four Sisters.

Although fragmented, the Kauri forests of Northland are important wildlife habitats. There are about 100 Kokakos in Puketi and Omahuta forests; along with the recently discovered northern Urewera Kokakos, they comprise probably the largest population of this endangered endemic bird now surviving in a single forest tract anywhere in the country. Brown Kiwi are found throughout the Kauri forests, which are also important habitats for Kaka, Red-crowned Parakeet, Kauri Snails and New Zealand's only indigenous terrestrial mammals, the Short-tailed and Long-tailed Bats.

The Waipoua River is the main waterway in the Waipoua Sanctuary Area, the strictly protected core of Waipoua Forest, New Zealand's largest remaining unmodified Kauri forest area.

LEFT New Zealand has 20 species of tree fern, so-called because the plant's rhizomes form a tall woody trunk topped by a crown of fronds. Mamaku (*Cyathea medullaris*), found also elsewhere in the Pacific, is one of New Zealand's tallest, reaching 20 metres (66 feet) and often emerging from the canopy; it is common in damp lowland forests (like Waipoua Forest) in the North Island and in coastal areas in the south of the country.

BELOW LEFT Tane Mahuta, New Zealand's largest remaining Kauri (*Agathis australis*). Like several other giants – including Yakas Kauri, Te Matua Ngahere and the Four Sisters – it is protected in the Waipoua Sanctuary Area. Once exploited for timber and Kauri gum but now widely regenerating, Kauri dominate many Northland and Coromandel forests. Their straight grey trunks lack branches until the spreading crown, which can form a canopy at about 35-40 metres (115-130 feet).

BELOW A resin, Kauri gum, can ooze from the trunks of Kauri trees and accumulate in the deep litter at the tree's base. Kauri gum was widely prized for a range of industrial uses, and from 1847 until the early part of this century large areas of forest were destroyed in the search for it. The 'bleeding' of trees for gum led to their weakening or death.

ABOVE A Paradise Shelduck (*Tadorna variegata*) chick foraging for food along the banks of the Waipoua River. This bird is widespread in forest waterways and open country.

LEFT The remaining Kokako (*Callaeas cinerea wilsoni*) populations are largely confined to a few podocarp/broadleaf forests in the central North Island and to Kauri forests like Puketi Forest in Northland.

BELOW Slow-growing lichens, like this foliose specimen, and moss species often appear as epiphytes on the trunks of older trees. Lichens are the product of a symbiotic relationship between a fungus and an alga (producing green lichens) or cyanobacterium (producing blue-green lichens). Blue-green lichens fix nitrogen, an important role in the nutrient cycle of the forest.

BELOW Kauri Snails (*Paryphanta busbyi*) feed mainly on worms. They move quite rapidly for snails, sometimes several hundred metres in a night. In the past they were often found living in the dense epiphytic vegetation of old Kauri trees but, with few of the Kauri surviving today, they now favour thick scrub and fern. Their heavy shells give them protection against rats and birds but wild pigs have ravaged colonies and destroyed their habitat.

BELOW *Gymnoplectron* cave wetas in Trounson Kauri Park. Cave wetas have longer legs than tree, ground and giant wetas, looking more like spiders. Although quite harmless, they tend to congregate in large groups on the roofs of caves and can be quite alarming when disturbed, jumping repeatedly. There are more than 50 cave weta species; the largest, like these, can have antennae 25 centimetres (ten inches) long and hind-legs 11 centimetres (4.3 inches) long.

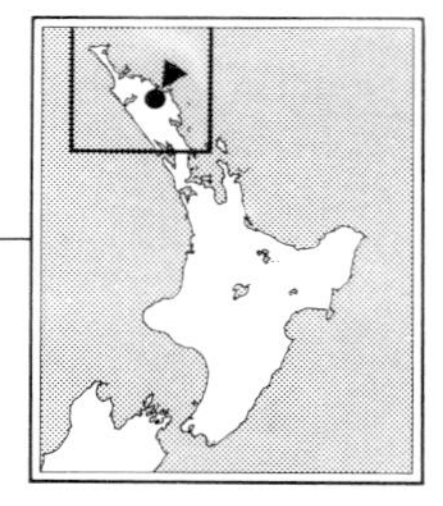

Bay of Islands

The Bay of Islands is often referred to as the 'birthplace of the New Zealand nation'. The locality supported a dense Maori population because of the shelter provided by the superb harbour and the abundant food supplies. The European navigators James Cook and Marion du Fresne, who came in the late eighteenth century, were soon followed by whalers, missionaries and traders. Today the Bay of Islands contains New Zealand's most important historic sites and buildings marking the earliest significant social and commercial contact between Maori and Pakeha – such as the early capital of Kororareka (modern-day Russell); Marsden Cross, site of the first Christian service for New Zealanders, both Maori and Pakeha; the Kerikeri stone store; and Waitangi National Reserve, site of the signing of the Treaty of Waitangi on February 6, 1840.

Many of the noteworthy historic sites and natural landscapes – including Urupukapuka Island and most of the major islands – in and around the inlets of this very extensive harbour are protected within the Bay of Islands Maritime and Historic Park. In addition, there is a comprehensive proposal to protect parts of the bay as marine reserves and close some other areas seasonally to fishing with nets and/or long lines. These widely supported conservation initiatives are designed to preserve representative examples of the diverse marine ecosystems and to protect key areas for recreational diving and rod-and-reel fishing. One of the most exciting sights in the park is the feeding frenzy of large predator fish like Marlin and Mako as they hunt schoolfish (Kahawai, Trevally or Mackerel). The threshing fish and foaming water attract further predators – gulls, terns, petrels and gannets – which all wheel about, skimming the surface or plunging into the water in search of their share of the feast.

Urupukapuka, the largest island in the Bay of Islands, has a mixture of Pohutukawa forest and farmland. It is popular for camping and diving.

ABOVE The estuaries of the Bay of Islands are fringed with groves of the mangrove *Avicennia resiniferra*, bringing a 'tropical' element to northern maritime landscapes and providing a habitat for a wide range of estuarine animals.

BELOW LEFT The hair-like plumage, cat-like whiskers and prominent ear-openings of the Brown Kiwi (*Apteryx australis*) are among the unusual features that set the Kiwi apart from other birds.

ABOVE Mainland Brown Teal (*Anas aucklandica chlorotis*) populations have been decimated since Europeans arrived. A small Bay of Islands population is being studied with the aim of re-establishing the species.

LEFT The endemic New Zealand Scaup (*Anthya novaeseelandiae*) is a small dark duck with great diving skills – even day-old chicks can dive. Adults can forage for food to depths over two metres (6½ feet) for up to 30 seconds.

Poor Knights Islands

The coastal aspect of the Poor Knights Islands is very impressive: sheer rhyolite cliffs rise for 150-200 metres (490–655 feet) above the waves and often also extend 100 metres (330 feet) below water level. But the summits of the islands are undulating and in places flat – the result of slow uplift leaving old beaches and wave-cut platforms as modern-day flat-topped ridges and terraces. The cliffs, both above and below water, are riddled with caves and archways which add to the scenic attractiveness and the diversity of the habitats in the islands.

Most of the original coastal forest was cleared off the two largest islands (Tawhiti Rahi and Aorangi) and the lesser islands in the group by the large Maori population which lived on the islands until they were declared *tapu* and abandoned after intertribal warfare in the early 1800s. Since then, forests of Kanuka, Pohutukawa and sometimes Tawapou have regenerated over the undulating plateaux, with Kohekohe, Karaka and Puriri in the gullies. The most notable plant is the beautiful Poor Knights Lily (*Xeronema callistemon*), which has a striking red flower and forms profuse carpets on rocky ground.

The islands are remarkable for their diverse and prolific wildlife, especially insects, reptiles and seabirds. Perhaps the most impressive insects are the Giant Weta and Cave Weta, the latter being of quite frightening dimensions – 30 centimetres (12 inches) from the claws on its hind-legs to the tips of its antennae. There are healthy numbers of Tuatara, geckos (two species) and skinks (five species); one of the geckos, Duvaucel's Gecko, is New Zealand's largest, reaching 30 centimetres (12 inches) in length. Because the islands are free of rats and cats, millions of seabirds are able to breed in safety. There are nine species of petrel, and these islands are the only breeding-ground for Buller's Shearwater, currently numbering between two and three million. The Poor Knights are among the most important wildlife-sanctuary islands in New Zealand.

RIGHT Archway Island, one of the smaller rocky islets in the Poor Knights group, is off the southern shore of Aorangi Island. The islands' sheerness can be seen also in the shapes of the two outlying islets, The Pinnacles and Sugarloaf, in the distance.

BELOW RIGHT Constant wave action has carved the rhyolitic rocks of the Poor Knights Islands into sheer cliffs, like these on Aorangi Island. Often the cliffs extend 100 metres (330 feet) below sea level, where underwater caves add to the diving attractions.

BELOW LEFT Most of the Poor Knights are very inaccessible, but Aorangi Island, one of two large islands in the group, has some shore platforms and tidal pools. The clear waters around the Poor Knights, an outstanding but vulnerable marine habitat, were protected in 1981 as New Zealand's second marine reserve.

LEFT The migratory Buller's Shearwater (*Puffinus bulleri*) is rarely seen on mainland New Zealand, but in late August over 200,000 return from the North Pacific to breed on the Poor Knights Islands. Pairs usually reclaim and vigorously defend the same burrow they used in previous years, although, surprisingly, some seem happy to share their subterranean nests with Tuatara – even though Tuatara are known to feed on Shearwater chicks!

The Shore Crab (*Leptograpsus variegatus*), common in mid-tide areas of exposed rocky coasts, can attain five centimetres (two inches) across its back.

The endemic Poor Knights Weta (*Deinacrida fallai*), slightly smaller than its Little Barrier relative but still up to seven centimetres (2¾ inches) long.

The Flax Snail *Placostylus hongii* is confined to coastal sites in eastern Northland. The shell can weigh about 17 grams (0.6 ounces).

This skink (*Cyclodina* spp) is found on the Poor Knights Islands. Most skinks are diurnal, but *Cyclodina* are most active at dusk and dawn.

The Pacific Gecko (*Hoplodactylus pacificus*), found in various North Island habitats from the coast to inland forests, has numerous varieties of colour and pattern.

New Zealand's largest gecko, Duvaucel's Gecko (*Hoplodactylus duvaucelii*), found only on a few offshore islands like the Poor Knights, is slow-growing and long-lived (up to 45 years).

ABOVE The Poor Knights Lily (*Xeronema callistemon*), when not flowering, looks much like a small flax bush. It can survive the toughest of conditions, with minimal soil and on windswept cliffs. Its only close relative, *Xeronema moorei*, grows in the mountains of New Caledonia.

ABOVE *Hebe bollonsii* is a coastal shrub about one metre (3¼ feet) tall, found only on the Poor Knights and Hen and Chickens islands. *Hebe* species show a remarkable ability to colonize a wide range of habitats, from rocky shoreline to subalpine tussock grasslands.

BELOW LEFT berries of the Pigeonwood or Porokaiwhiri (*Hedycarya arborea*). As the names suggest, the berries are a favourite food of the New Zealand Pigeon. Pigeonwood is found throughout the North Island and as far south as Banks Peninsula.

BELOW The rare Poor Knights Lily, found only on the Poor Knights and on Taranga Island (in the Hen and Chicken group). The flower has no petals, its colour and striking appearance arising from the very visible stamens. The flowers mature into dry brown capsular fruit.

AUCKLAND, COROMANDEL AND THE ISLANDS OF THE GULF

New Zealand's largest city – Auckland – straddles a narrow volcanic isthmus linking Northland with the rest of the North Island. To the east lies the Waitemata Harbour, with the island-studded Hauraki Gulf beyond; to the west lie the Manukau Harbour and the Waitakere Ranges.

The differences in coastal landscapes between east and west are striking. The waters and islands of the gulf are sheltered by the long arm of the Coromandel Peninsula and Great Barrier Island; the coastline from Cape Rodney to the Firth of Thames is deeply indented, with many rocky headlands and peninsulas, mangrove-fringed estuaries and small sandy beaches. In contrast, the rugged rocky coastline and isolated beaches west of the Waitakeres are continuously pounded by surf driven by the westerly winds which sweep in from the Tasman Sea.

Most visitors to Auckland do not immediately realize that the city is built on a field of volcanoes. Over the past 60,000 years more than 50 volcanoes have built up the Auckland Isthmus, but many of these small basaltic cones have been quarried away or partly covered in houses. All of them are now extinct or dormant, although future volcanic activity is likely. Many of the cones are now prominent city landmarks (One Tree Hill, Mount Eden, North Head and Mangere Mountain, for example). The youngest and largest of the volcanoes, the island of Rangitoto, is the least modified and the most graceful in profile.

The Coromandel Peninsula is the nearest wild mountainous landscape to Auckland, lying only about 65 kilometres (40 miles) away across the Hauraki Gulf and Firth of Thames. Most of the peninsula consists of very old andesitic volcanic rocks, now deeply eroded to leave a jagged maze of bluffs and pinnacles. The crest of the Coromandel Range is one of the most dramatic skylines in the North Island, extending 110 kilometres (68 miles) from Cape Colville at the northern tip to, in the south, Karangahake Gorge, the site of the largest complex of historic mining sites in the North Island. The history of the Coromandel has largely been one of exploitation – for gold and silver, and the golden timber of the Kauri. The forests of the Coromandel Range no longer extend down to the beautiful coastline: the lowland forests were cleared long ago in largely futile attempts to farm the infertile clay soils.

The Coromandel coastline has much to offer the visitor. The beaches are some of the best in the country, many lying in secluded bays, far from the main tourist routes but with simple campgrounds nearby. Three of the more isolated east-coast beaches still retain a high proportion of their original sand-dune vegetation: Waikawau, Otama and Hot Water Beach, the latter so-called because of thermal waters percolating up through its sands, providing a delightful natural spa-like diversion for the weary visitor. One of the most attractive stretches of coastline and seabed, lying between Cathedral Cove and Hahei Beach, is now the Te Wanganui a Hei Marine Reserve.

Great Barrier Island (or Aotea) is – at 285 square kilometres (110 square miles) – the largest of the islands off the coast of the North Island. It lies only 20 kilometres (12.5 miles) across the Colville Channel from the northernmost tip of the Coromandel Peninsula, to which it was joined during the last glaciation, when sea levels were much lower than they are today. Like the peninsula, it is mainly of volcanic origin and offers an extremely rugged forested landscape. Although the island was ruthlessly stripped of its Kauri forest during the late nineteenth and early twentieth centuries, there is now widespread regeneration of the Kauri forests. Many relics of this exploitative colonial economy can still be seen; the Kauri dams and tramways, in particular, are major visitor attractions and are considered the best in the country.

Despite its history of habitat destruction, the conservation values of Great Barrier are of national significance, and over 60 per cent of the island is now protected as some form of conservation land. The track from 'Windy Canyon' across the 621-metre (2,036-foot) summit of Hirakimata (Mount Hobson) and down the steep western slopes into the Kaiarara Valley is one of the region's outstanding recreational experiences. Features include spectacular canyon landforms, bewildering botany, seabirds that nest in burrows on mountain tops, and well preserved Kauri dams that evoke a bygone era now being erased by the profusely regenerating forest.

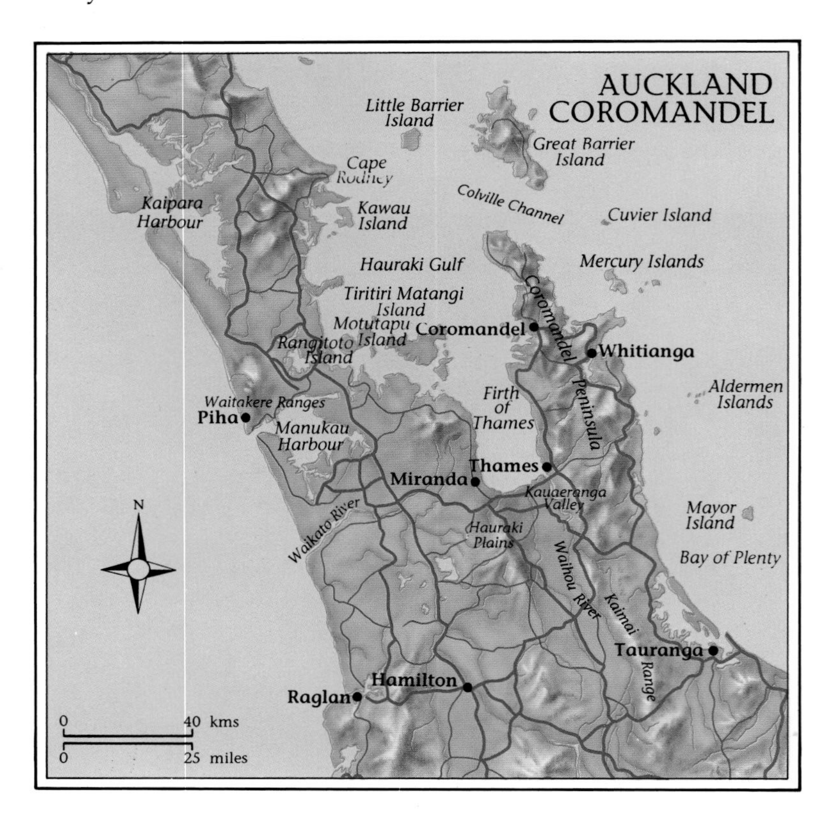

RIGHT The Coromandel Peninsula's coastline is a delightful mosaic of rocky headlands and bays, with many small islands offshore. Cathedral Cove, near Hahei, is now part of the Te Wanganui a Hei Marine Reserve.

The island is the largest area in New Zealand still free of three key groups of wild introduced animals – possums, mustelids and deer. There are some notable native wildlife populations: about 60 per cent of the remaining Brown Teal in the country; important Kaka, Spotless Crake, Banded Rail, New Zealand Dotterel and Fernbird populations; the principal nesting sites for Black Petrel in New Zealand; the only known island population of the endangered Hochstetter's Frog; and one of the most diverse lizard faunas in the country, with at least 13 species, including the Great Barrier Skink.

Within the sheltering arm of the Coromandel and the Great Barrier (aptly named by Captain James Cook on his first voyage to New Zealand) lie the islands of the Hauraki Gulf. These fall into two broad groups: the outer islands, like Little Barrier, Cuvier and the Hen and Chickens, are wildlife sanctuaries; the inner islands, like Kawau, Rangitoto and Motutapu, are popular for recreation. Public access to the sanctuary islands is carefully controlled to avoid the inadvertent introduction of further exotic pests that might threaten the indigenous wildlife.

Little Barrier Island lacks browsing mammals, Ship Rats and Norway Rats; its relative freedom from introduced mammals is typical of most of the outer-island sanctuaries. The Kiore (the Pacific Rat) is present, and probably accounts for the declining numbers of Tuatara and Giant Weta.

The notable conservation achievement on Little Barrier has been the elimination of cats – the result of an internationally acclaimed campaign intensively conducted between 1977 and 1980 by the New Zealand Wildlife Service with the aid of volunteers. The effect on some bird populations was amazing. One of the rarest seabirds, the Black Petrel, which nested only on the summit peaks, was on the verge of elimination, while the beautiful Tieke (the North Island Saddleback, a wattlebird in the same family as the Kokako and Huia) had been eliminated long before by the cats because of its ground-nesting habit. New Zealand's rare honeyeater, the Stitchbird, had also been almost wiped out, but its numbers exploded tenfold to 3,000 in the decade following the early 1970s, when the losses to predation by cats had been at their peak – or, looked at another way, their nadir.

Without cats the island became safer, too, for the introduction of the most vulnerable of New Zealand's birds: first the Kokako, rescued from native forests being felled in the Rotorua district, then small numbers (ultimately 20) of the rarest parrot, the Kakapo, rescued in the nick of time from cats on Stewart Island. Saddlebacks were reintroduced, robin numbers bounced back and seabird populations slowly responded.

Two parts of the eastern coastline of the Auckland region have particular wildlife significance. The first is an extensive area of tidal flats at the southernmost extent of the Firth of Thames, itself a southern arm of the Hauraki Gulf. Between Miranda and Thames this 'wetland of international significance' supports up to 40,000 migratory wading birds including Eastern Bar-tailed Godwit, Sharp-tailed Sandpiper, Mongolian Dotterel and Ringed Plover, as well as New Zealand's own unique Wrybill.

The other protected coastline is from Cape Rodney to Okakari Point near Leigh, 60 kilometres (37 miles) north of Auckland city. In 1975, 574 hectares (18 square miles) of this geologically diverse piece of seabed and coastline were protected as New Zealand's first marine reserve. The marine environment of the reserve is largely clean and silt-free, in contrast to the rest of the inner Gulf area. The changes in the richness of the inshore and reef-dwelling marine life with the cessation of fishing have been scientifically monitored from the adjacent Auckland University Marine Laboratory. Dramatic improvements in fish size and species diversity are partly responsible for the fact that the reserve now attracts about 100,000 visitors per year. Its overall success has been largely due to the vision of Dr W.J. Ballantine and his university colleagues, who have helped raise public awareness of marine conservation nationally, so that the marine reserve has become a maritime 'conservation milestone'. Public support for the concept is increasing, and a comprehensive network of marine reserves began to appear in the early 1990s after an unfortunate hiatus of 15 years.

RIGHT North of Manukau Harbour and Whatipu, Auckland's western coastline is a series of rocky shelves and cliffs sandwiched between the Waitakere Ranges and the sea.

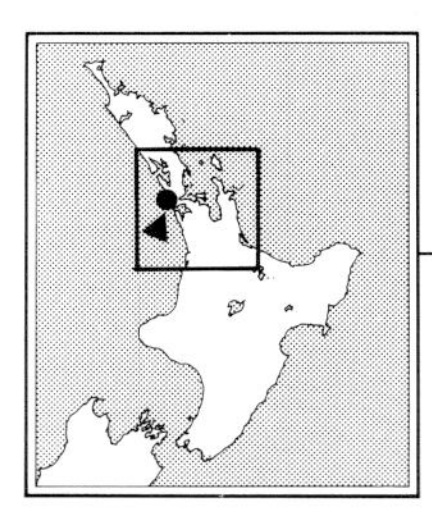

Auckland's Wild West Coast

The Waitakere Ranges on the western skyline of the city are a popular forest retreat for Aucklanders. Although there are no major rivers, there is an intricate network of streams and steep ridges with many small waterfalls. The mixed forest now shows little visual evidence of its past history of logging, although travel is difficult through the areas of regenerated cut-over forest. Kauri are everywhere regenerating and beginning to penetrate the predominantly Tawa forest canopy. In addition, isolated large Rata, Rimu, Totara and Kahikatea tower above the lesser canopy trees. The moist gullies have a tropical-rainforest feeling, with their profusion of Nikau Palms, tree ferns and, hanging like ornamental baskets from the trees, countless epiphytes. The visitor centre operated by the Auckland Regional Council at Arataki is the focal point for information on walking and wildlife-observation opportunities.

The wild western coastline can be reached by road in a number of places: Whatipu, at the entrance to the Manukau Harbour, is one of the most impressive dune landscapes in New Zealand, while further north lie Karekare, Piha, Te Henga and Muriwai. Each of these beach localities has its own untamed character and sense of danger, and each is popular with surfers, surfcasters and lovers of wildlife and wild coastal scenery. The lava flows and dykes in prominent landmarks like Nun Rock and Lion Rock have been carved by the pounding seas into high narrow caves and chimneys, which become fuming blowholes in high seas. The water temperatures are three to four Celsius degrees (5–7 Fahrenheit degrees) lower than in the Hauraki Gulf on the eastern side of the isthmus, and the marine life is more typical of New Zealand's colder coasts (for example, the massive Bull Kelp *Durvillea antarctica*). On spectacular coastal stacks at Muriwai the visitor can easily observe one of only three mainland breeding colonies of the Australasian Gannet (the other two sites are Cape Kidnappers, at the southern end of Hawke's Bay, and Farewell Spit – see page 124).

The rocky headland of Muriwai Beach hosts the second largest mainland colony of Australasian Gannet (*Morus serrator*). In general, gannets mate for life, and their greeting protocols and courtship displays appear to be part of the strict order in the bustling colony.

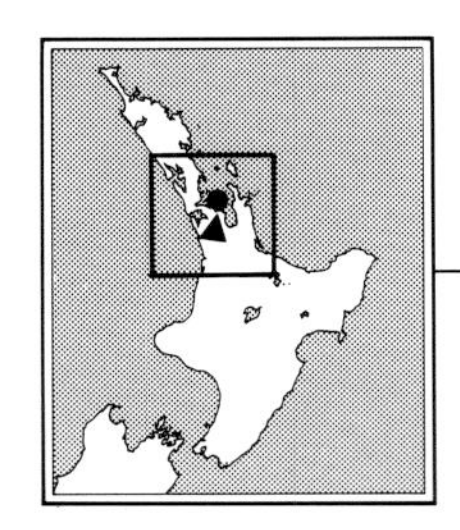

Rangitoto Island

Rangitoto Island is a perfect example of a shield volcano. Over the last 800 years its gently sloping symmetrical shape has built up from a series of basaltic lava eruptions; the lava was mobile enough to flow in all directions, thus creating a circular island six kilometres (3.7 miles) in diameter. The steeper-sloped summit cinder cone is probably younger, representing the accumulation of coarse airfall particles (scoria) from a localized eruption. To those familiar with the great shield volcanoes of Mauna Loa and Mauna Kea on the island of Hawaii, Rangitoto seems a superb replica, in miniature.

The similarity of Rangitoto to Hawaii lies not only in its shape but also in its vegetation. The lava-fields of both islands are colonized by closely related species of *Metrosideros* – in New Zealand known as the coastal tree Pohutukawa, in Hawaii as the Ohia. The first impression on approaching Rangitoto across the Waitemata Harbour is of the verdant forests clothing its graceful slopes. The reality, as you discover on disembarking, is quite different, for there is no running water, and blisteringly hot black-lava flows extend their long fingers through the thin Pohutukawa forest. The young lava surface is almost devoid of mineral soil, but leaves accumulate at the base of isolated Pohutukawa trees and rapidly build up a humus layer.

Because of the youthful nature of the island and the lack of soil development, the vegetation of Rangitoto is almost unique in New Zealand; the only similar community is on the tholoid (dome) in the interior of Mayor Island in the Bay of Plenty. Rangitoto is an outstanding place to study the intricacies of plant colonization. In less than 800 years the rough aa (block) lava of a new island has been colonized by more than 200 species of trees and flowering plants and over 40 different ferns – including entrancing groves of the translucent Kidney Fern.

Rangitoto is the scenic gem of the Hauraki Gulf, so accessible from downtown Auckland yet a world apart in its austerity and simple beauty.

Nowhere are organized settlement and unpredictable nature better contrasted than in Auckland. The dormant volcano Rangitoto Island lies only three kilometres (two miles) offshore.

ABOVE Pohutukawa (*Metrosideros excelsa*) clings to inhospitable sites using long twisted roots and tangled, fibrous, aerial roots. Its flowers decorate northern cliffs and coastal forest in summer – hence the nickname 'New Zealand Christmas tree'.

ABOVE Kidney Ferns (*Trichomanes reniforme*) form dense mats in a wide range of forest habitats. They prefer moist conditions; on Rangitoto's dry lava-fields they seek sheltered sites under canopies.

BELOW The hardy white moss *Racomitrium lanuginosum* is often first to colonize rocky sites, from Rangitoto's lava-fields to alpine screes.

BELOW Temperature fluctuation above the bare, black basaltic lava-flows of Rangitoto is extreme. Mosses colonizing exposed rocky surfaces help regulate the temperature; thereafter seedlings more easily establish.

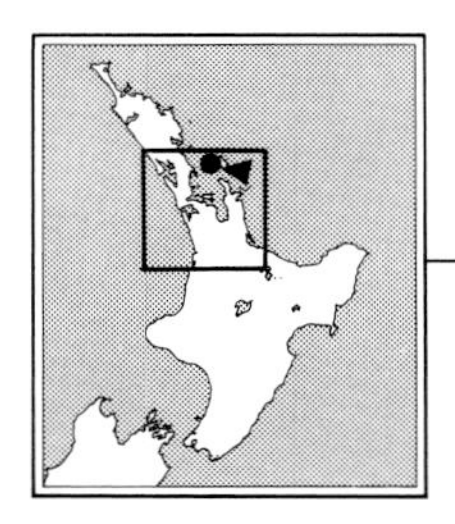

Little Barrier Island

Little Barrier is by far the most important of the sanctuary islands of the Hauraki Gulf. A large island – 3,083 hectares (11.9 square miles) – it is the deeply eroded cone of a roughly circular extinct andesitic volcano. Because of its steep slopes and height – 722 metres (2,367 feet) – it is a dramatic part of the seascape seen from the nearest part of the mainland, 22 kilometres (13.7 miles) away at Cape Rodney. The Maori name for this impressive island is Hauturu ('The Wind's Resting Post'), a name bestowed by the explorer Toi on his arrival in Aotearoa about 850 years ago.

It is now a century since the New Zealand Government purchased Hauturu from its Maori owners in recognition of its potential importance as a wildlife sanctuary. Densely forested, the island boasts the best remaining altitudinal sequence of indigenous vegetation in northern New Zealand. The coastal forest consists of gnarled old Pohutukawa joined upslope by Kauri, Hard Beech, Rata and Tawa. At altitudes of about 500 metres (1,640 feet), Tawa and Tawhero form an amazingly tight high-humidity rainforest, their trunks and branches festooned with filmy ferns and mosses often right up to the canopy. Finally, on the most windswept ridges above 650 metres (2,130 feet), lie a tangled stunted forest and shrubland dominated by three hardy montane trees – Quintinia, Tawari and Southern Rata. The diversity of plant-life is one of the highest in the country – over 370 higher plants, including 90 different varieties of fern. Interestingly, the podocarps, usually such dominant trees in the New Zealand forest, are, save for isolated Miro and Hall's Totara, largely lacking. Even the elsewhere ubiquitous Rimu is completely absent, a circumstance that poses a floristic puzzle.

Today Little Barrier Island is a priceless sanctuary, strictly protected as a Nature Reserve. It plays a key role in the recovery of New Zealand's endangered birds, like the Kakapo, amply justifying the faith of those far-sighted individuals who, well over 100 years ago, began the campaign for its acquisition.

Little Barrier Island (Hauturu) has one of New Zealand's best altitudinal sequences of forest vegetation, from sea level to windswept spurs above 650 metres (2,130 feet).

ABOVE The New Zealand Pigeon (*Hemiphaga novaeseelandiae*) is an obvious bird in podocarp/broadleaf forests because of its large size, conspicuous markings and noisy flight.

BELOW The large Chevron Skink (*Leiolopisma homalonotum*), so-named for the distinctive markings down its back and tail, is among New Zealand's rarest skinks, now found only on Great Barrier and Little Barrier islands.

ABOVE The Nikau Palm (*Rhopalostylis sapida*), New Zealand's only member of the Palm family, found on Little Barrier Island and throughout the Hauraki Gulf, is a common coastal subcanopy tree in the milder parts of the country.

ABOVE The Blue Penguin (*Eudyptula minor*), the smallest of all penguins, nests widely around mainland New Zealand and on offshore islands like Little Barrier. Mainland populations are threatened by dogs, ferrets and, on coastal roads, vehicles.

ABOVE RIGHT The beautiful Tieke or North Island Saddleback (*Philesturnus carunculatus rufusater*), one of New Zealand's surviving wattle-birds (along with the Kokako). Predators eliminated it from the North Island, and for many years it was confined to Taranga (Hen) Island until being transferred to other safe islands.

BELOW LEFT Kohurangi (*Brachyglottis kirkii*), found in lowland or montane forest throughout the North Island as either a ground-dwelling shrub or an epiphyte. Ravaged by the introduced Australian Brush-tailed Possum, it is increasingly rare.

RIGHT The bark of Kauri trees (*Agathis australis*) regularly peels off to leave a distinctive 'hammer-strike' pattern. This may be a mechanism to prevent epiphytic plants from establishing or persisting.

BELOW CENTRE *Hebe macrocarpa latisepala*, found in coastal forest and scrub on Great and Little Barrier islands, has a deeper purple flower than mainland varieties and, unlike those, retains the colour as the flower ages.

BELOW RIGHT A coral fungus, probably a species of *Ramaria* in the Clavariaceae family, growing on rotting wood on a forest floor in Little Barrier Island.

Tiritiri Matangi Island

The island of Tiritiri Matangi, of area 220 hectares (543 acres), is one of the Hauraki Gulf's most interesting. It is managed by the Department of Conservation as an 'open sanctuary', where visitors are encouraged and assisted not only to learn about their natural heritage but also to play an active role in improving it. Tiri, as it is usually called, has been a 'lighthouse island' for 130 years, and was grazed for over a century. The grand plan is to rehabilitate the former forest cover of the island through volunteers replanting vegetation. Like most other islands in the gulf, Tiri had a long history of Maori occupation, evidenced by the remains of middens and characteristic kumara-storage pits and by the surviving population of Kiore (the Pacific Rat), which is in the process of being eradicated from the island.

The conservation effort to date has been impressive. In the past decade the Supporters of Tiritiri organization and other volunteers have planted over a quarter of a million trees. The pioneer tree plantings were of the hardy Pohutukawa, followed by more shade-requiring species like Kohekohe, Puriri and Taraire. The Department of Conservation has continued the bird-reintroduction work initiated by the Wildlife Service and Ducks Unlimited. The first birds released were red-crowned parakeets, Saddlebacks, Whiteheads and Brown Teal. More recently, two of New Zealand's rarest birds – the Little Spotted Kiwi and Takahe – have been liberated on the island and appear to be thriving.

Tiritiri Matangi has had an interesting history: Maori occupation, farm, lighthouse reserve and now an 'open sanctuary' for endangered wildlife.

LEFT The North Island Fantail (*Rhipidura fuliginosa placabilis*) is among the most appealing forest birds, performing marvellous feats of flight in the pursuit of insects. It is at home in the forest, on farmland or in the suburban garden, having adapted well to the changes wrought by European settlement. The species has a dark colour variant as well as the 'normal' pied form.

BELOW LEFT Grey-faced Petrels (*Pterodroma macroptera gouldi*) breed in winter, laying a single egg in June or July; both birds incubate the egg, in spells averaging 17 days! To minimize energy-loss the incubating bird spends most of the time in deep sleep. Fledglings were once widely harvested, but most colonies are now protected.

BELOW The rapid decline of the once widely distributed Brown Teal (*Anas aucklandica chlorotis*) prompted their total protection in 1921, but this has not halted their diminution due to predation and habitat loss. They have been introduced to Tiritiri Matangi, but the largest populations remain in Northland and on Great Barrier Island, where they live in wetlands, ponds and swampy pasture.

Coromandel Peninsula

Despite the past logging of the Kauri forests of the Coromandel Peninsula, there are still isolated patches of mature Kauri forest in places like the Manaia Sanctuary and the Kauaeranga Valley in the heart of the rugged Coromandel Forest Park. There is a comprehensive network of tracks leading from the visitor centre in the Kauaeranga Valley; many of them lead to historic Kauri logging tramways or dams built to hold back the head of streamwater for the notoriously destructive Kauri log-drives.

At the northern end of the park, like a forested island rising sheer from the sea, lies the Moehau Range. Here, around Mount Moehau – at 892 metres (2,924 feet) the highest point on the peninsula – lies the northernmost subalpine locality in New Zealand. Boggy herb-fields and shrublands on the undulating crest contain a great number of plants far more characteristic of the greywacke ranges at the southern end of the North Island. The Moehau summits are sacred to many Maori as the traditional resting-place of Tama te Kapua, leader of the Arawa canoe in its migration from Hawaiki to Aotearoa.

Possums have still not penetrated the Kauri/broadleaf forests of the Moehau Range, and the Department of Conservation has made an interesting proposal for protecting this northern forest 'island': the erection of a 'possum-proof' fence across the narrow neck between Colville and Waikawau Bays. The Coromandel forests have some outstanding wildlife features: there are Kaka, Red-crowned Parakeet, New Zealand Falcon, Brown Kiwi and isolated small populations of Kokako. The forests are also noted for their less conspicuous fauna: the rare Coromandel Stag Beetle and two species of New Zealand's primitive native frogs – Hochstetter's Frog and Archey's Frog. The latter, very rare, is found only high up in the Coromandel Range, in forest litter and usually far from running or standing water.

The Coromandel Peninsula is a complex of old volcanic rocks, often eroded by the sea into arches like this one in Mercury Bay.

ABOVE LEFT The Coromandel Forest Park occupies 75,000 hectares (290 square miles) of the Coromandel and Moehau ranges, mostly heavily forested except where lava-flows, dikes and sills have produced a skyline of sheer-sided pinnacles and domes.

BELOW LEFT The tallest tree ferns, Mamaku and Punui (both *Cyathea*), reach 20 metres (66 feet) and are widespread on all main islands from sea level to 800 metres (2,625 feet).

BELOW CENTRE Common throughout the country, this robust, unnamed fern of the *Blechnum* genus requires good light and grows mainly on moist, steep-sided banks and gorges.

ABOVE RIGHT New Zealand Dotterel (*Charadrius obscurus*) vigorously defend their breeding territories. The birds may feign injury to draw persistent intruders away from the nest.

RIGHT Hochstetter's Frog (*Leiopelma hochstetteri*), the commonest native frog, lives in northern North Island. It has no free-swimming tadpole stage but, unlike Archey's and Hamilton's frogs, lives and lays its eggs near water.

BELOW RIGHT Smallest of the four native frogs, Archey's Frog (*Leiopelma archeyi*) lives only in the bush-covered hills of the Coromandel and western King Country, thriving in moist vegetation but often far from running water.

Mercury Islands

The Mercury Islands are a group of seven islands lying 7–15 kilometres (4.3–9.3 miles) off the eastern coastline of the Coromandel Peninsula; they have considerable wildlife significance. All volcanic in origin, they were joined to the peninsula during the last ice age, but were separated by rising sea levels about 9,000 years ago.

Great Mercury Island is privately owned, but Red Mercury Island and the five small islands in the group are strictly protected as nature reserves. Despite past Maori occupation, fires and the introduction of Kiore and rabbits to some of these islands, their fauna still retains many interesting features. Vast numbers of burrowing petrels on the smaller rat-free islands have markedly increased the fertility and aeration of the soils. These conditions seem to suit rapid plant growth and the production of litter and topsoils that support a dense and diverse fauna of invertebrates and lizards, as well as some unusual Milk Tree and Wharangi-Mahoe forests.

These small island ecosystems give an excellent insight into what New Zealand may have been like before introduced mammals displaced the indigenous non-mammalian predators. For example, the small Middle Island – only 13 hectares (32 acres) in area – has the unique, formidable-looking Tusked Giant Weta which, unlike the other Giant Wetas and Tree Wetas, is omnivorous. There are ten species of lizard on Middle Island (three geckos and seven skinks, including the very rare Whitaker's Skink and Robust Skink), making this tiny island's lizard population one of the most diverse in the country. Another resident, found in appreciable numbers only on northern rat-free islands, is the large carnivorous centipede *Cormocephalus rubriceps*, which is capable of feeding even on lizards. Finally, there is the top predator, the Tuatara, present in greatest numbers on the small rat-free islands like Middle Island; on the larger islands of Red Mercury and Stanley it was, by contrast, until the recent eradication there of Kiore, on the verge of extinction.

Red Mercury Island, the second largest of the seven Mercury Islands, is named for the colour of the basalt scoria in its cliffs.

ABOVE The Aldermen Islands, 20 kilometres (12½ miles) off the eastern Coromandel coastline, are the remnants of a rhyolitic volcano. Although only 134 hectares (330 acres) in area, they are an important Tuatara habitat.

RIGHT The Tusked Weta of tiny Kiore-free Middle Island is one of the rarest and most formidable-looking of New Zealand's wetas. The males have large tusks growing from their jaws, which they use in territorial disputes with other males. With their long legs they can jump more than a metre (40 inches)!

BELOW LEFT The Giant Centipede (*Cormocephalus rubriceps*), 20 centimetres (eight inches) long, with a venomous bite and spines on its hind-legs to hold its prey, is now found only on northern rat-free islands.

BELOW RIGHT Tuatara are populous on some smaller Kiore-free islands in the Mercury group, and numbers are recovering on Red Mercury and Stanley islands since Kiore eradication.

THE VOLCANIC LANDS AND FORESTS OF THE CENTRAL NORTH ISLAND

The central North Island region has a character that sets it apart from the northern areas – with their lowlands, coastal influences and warm temperate soils and vegetation. Here, by contrast, mountains dominate the skyline, the climate is distinctly cooler and wetter and the influence of volcanism is all-pervasive.

At the very centre of this region lie Lake Taupo and the so-called Volcanic Plateau, a wild and highly scenic landscape of volcanoes, domes, lake-filled calderas and high lava plateaux that slope down to the Bay of Plenty and the Waikato Basin. The first European settlers who tried to establish farms in this difficult country called them the 'pumice lands'. Pumice, a distinctive soft, greyish-white volcanic rock, light enough to float in water, mantles the surface of most of these landforms as a result of the great Taupo eruption 1,800 years ago. The main wild rivers of the North Island – the Waikato and its major tributary, the Waipa, the Rangitaiki, Mohaka, Ngaururoro, Rangitikei and Whanganui – radiate in all directions from the Volcanic Plateau and provide important freshwater habitats.

Encircling the Volcanic Plateau is a series of very rugged forested ranges, which together make up the forest wilderness core of the North Island. To the east lie the axial greywacke ranges – Ruahine, Kaimanawa, Kaweka and Huiarau – the 'backbone' of Te Ika a Maui ('The Fish of Maui'). To the west lie the Mamaku Plateau and the Hauhungaroa and Matemateaonga ranges. The latter links the Volcanic Plateau with the volcanic outlier of Mount Taranaki by way of the vast sedimentary basin of the Whanganui River. The basement rocks of these perimeter ranges are not volcanic in origin, but everywhere their slopes were in the past mantled with tephra (airfall ash and gravel from volcanic eruptions); although in many places this has largely been eroded away, rivers like the Whanganui still transport floating pieces of pumice and other volcanic debris to the western coastline. Indeed, the iron-rich sands of the North Island's west-coast beaches originated in the volcanic rocks (particularly the andesites) of the Volcanic Plateau and Taranaki.

Lake Taupo and the Rotorua lakes lie in a rift valley which geologists call the Taupo Volcanic Zone. This zone of weakness in the Earth's crust extends in a striking line of volcanic landscapes from White Island in the Bay of Plenty for 250 kilometres (155 miles) to the Ohakune Craters on the south-west flanks of Mount Ruapehu. The Taupo Volcanic Zone can be thought of as a southern terminus of the great Pacific Ring of Fire which extends down the western Pacific from the Tongan Islands through the Kermadec Islands (see page 50) to the Bay of Plenty. The zone contains, in White Island and Tongariro National Park, some of the most continuously active volcanoes in the world.

To the Maori there are also unity, awe and danger in the volcanic

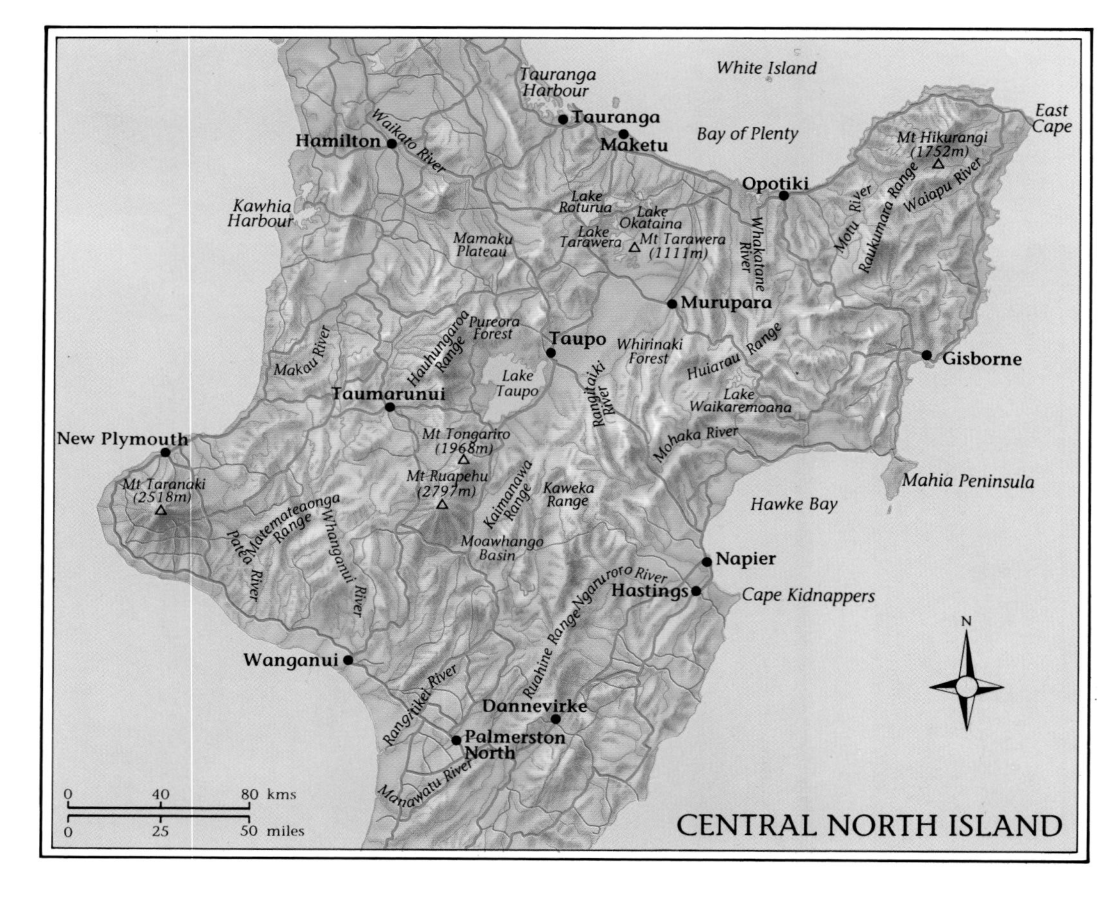

OPPOSITE The active volcano of Mount Ngauruhoe, 2,287 metres (7,503 feet) tall, is a prominent landmark in Tongariro National Park. The park lies at the southern end of the Taupo Volcanic Zone, the great volcanic rift through the central North Island. This winter view is from the Rangipo Desert, a harsh environment of subalpine gravel-fields and Red Tussock grasslands on the park's eastern margin.

landscapes of the Bay of Plenty and the central plateau. The area is peopled by the descendants of the Arawa canoe that made its final landfall from Hawaiki at Maketu in the Bay of Plenty. The navigator and *tohunga* of the Arawa, Ngatoroirangi, daringly explored the volcanic hinterland, finally climbing the snowy summit of a high mountain. Close to death from the icy wind (hence the name Tongariro, 'carried away by the cold south wind'), he called in prayer on his sisters in Hawaiki to send fire to warm him. The fire gods obliged, and it came in a straight line, bursting forth in lava and steam along what we now know as the edge of the Pacific Plate – Tongatapu, Rangitahua (Raoul Island in the Kermadec archipelago), Whakaari (White Island), Rotoehu, Rotorua, Whakarewarewa, Tarawera, Orakei Korako, Wairakei, Taupo, Tokaanu and, finally, Tongariro.

Vulcanologists believe that activity began in the rift about two million years ago; the oldest volcanoes, Titiraupenga and Pureora in Pureora Forest Park along the western margin, date from this time. The most widespread landforms, however, are the huge plateaux consisting of ignimbrite – a rock 'welded together' from molten rhyolite lava and volcanic gas ejected at very high temperatures. These plateaux were laid down 750,000 to 300,000 years ago. In the east, the vast Kaingaroa Plateau slopes gently down from Lake Taupo and today is mainly covered with plantation forests of introduced conifers; in the west, the plateaux

– Mamaku, Tokoroa and West Taupo – are, though smaller and less uniform, of high conservation value because they still carry podocarp forests that are very important wildlife habitats; for example, Rotoehu, Mamaku and Pureora forests.

Tongariro National Park protects the complex of andesitic volcanoes at the southern end of the Taupo Volcanic Zone. The skyline is dominated by the huge massif of Ruapehu, the southernmost of the seven main volcanic peaks in the park and, at 2,797 metres (9,170 feet), the highest mountain in the North Island. The isolated vent of dormant Hauhungatahi lies to the west, while the massif of Tongariro (and its classic daughter cone, Ngauruhoe) stands 15–20 kilometres (9–12 miles) to the north-east. At the northern extremity of the park three beautiful forested volcanoes – Pihanga, Tihia and Kakaramea – rise above the southern shores of Lake Taupo.

Despite the symmetrical simplicity of Ngauruhoe, there is nothing simple about the history of these volcanoes. Ngauruhoe is really just the most recent (only 2,500 years old) and most active of at least a dozen cones, vents and craters that go to make up the sprawling landscape of a single 'Tongariro volcano'. Ruapehu is likewise active and complex in origin, with five craters and six main peaks. In addition, both massifs were sculpted by glaciers during the last ice age, and even today small remnant glaciers – the most northerly in New Zealand – cling to the flanks of Ruapehu.

Maori oral history contains a graphic account of the relationships between these volcanoes in what eventually became known as Tongariro National Park and those elsewhere in the central North Island. Mountains were believed to travel, and sometimes quarrel. The male volcanoes of the Taupo Volcanic Zone fell out over the beautiful female volcano Pihanga, whose most powerful and persuasive suitor was Tongariro. His wrath towards her other suitors exploded, and the vanquished volcanoes scattered in several directions. His chief rival, Taranaki, fled by night to the west, carving out the trough of the Whanganui River before travelling up the coast to a safe home in the region that now bears his name. Putauaki (Mount Edgecumbe) travelled as far as the north-eastern end of the rift valley. Tauhara, who could not bear to lose sight of Pihanga, went only to the northern shores of Lake Taupo, where he stands as a lonely sentinel, gazing longingly across the waters to where his loved one looms above the southern shore.

Mounts Ruapehu and Ngauruhoe are two of the most continuously active composite volcanoes in the world. Composite volcanoes are sometimes called stratified volcanoes (or stratovolcanoes) because their steep sides are built up from layers of lava and airfall material (ash, gravel and boulders), collectively called tephra. Other well known composite volcanoes, like Fujiyama and Kilimanjaro, have not erupted in the last century, but during the same period Ruapehu has given vent to explosive eruptions on average more frequently than once per year. In addition, the hot-water Crater Lake, occupying the summit of Ruapehu, is unique because of its glacial setting and the violent steam eruptions caused when rising magma comes into contact with the lake waters. On Christmas Eve, 1953, the collapse of the ice barrier impounding Crater Lake gave rise to a lahar (mud-lava flow) that swept away the Wellington-Auckland express train and the Tangiwai railway bridge 30 kilometres (19 miles) away, with the loss of 151 lives in what is still New Zealand's worst railway disaster.

The volcano that formed Lake Taupo has been active for the last 330,000 years; it is reputed to be the most frequently active ejecta-producing rhyolitic volcano in the world. For probably the last 100,000 years a lake has occupied its caldera. The mixing of the lake water with magma rising in the vent causes the extraordinarily powerful steam explosions that have characterized Taupo's eruptions. The most recent of these, recorded in Chinese and Roman annals for AD186, though not the largest in the volcano's history, was nevertheless the most violent anywhere in the world in the last 5,000 years. It pulverized some of the ejected material to fine ash, which spread across most of central and eastern North Island, up to three metres (ten feet) thick along the eastern shores of the lake and still ten centimetres (four inches) thick as far away as Gisborne. But the culmination of the eruption was an awesome pyroclastic flow of incandescent gas and volcanic debris, so fluid and frictionless that its horizontal speed was close to that of sound. It rode over all the volcanic peaks except the summit of Ruapehu, instantly flattening forests and incinerating them and their wildlife. It is hard to visualize the release of so much energy almost instantaneously; it is estimated that in just a few minutes a volume of 30 cubic kilometres (7.2 cubic miles) of volcanic material was ejected from the vent.

About 75 kilometres (47 miles) along the rift to the north-east lies the squat rhyolitic dome of Mount Tarawera, with a line of blood-red chasms slashed along its broad summit. Here, close to the present-day tourist resort city of Rotorua, occurred New Zealand's most violent volcanic eruption of the past 150 years. In the early hours of June 10, 1886, Tarawera erupted without warning. The eruption lasted only four hours, but had devastating human and environmental consequences, with over 150 people being killed.

Most of the casualties were Maori in the local villages, many of whom had served as guides for the well established tourist industry around Tarawera. Visitors had come to Rotorua from all around the world to see the fabled Pink and White Terraces, which cascaded down the slopes above what was then a much smaller Lake Rotomahana and were reputed to be the best geothermal sinter terraces in the world. These were completely destroyed by the eruption, and in their place developed a much larger Lake Rotomahana, which now received the hot waters of the Waimangu Valley – a new, extremely violent, geothermal locality.

The Volcanic Plateau must always have been a harsh environment for native plants and wildlife. Previously much of it was covered in a heathland of Manuka and Monoao. The combined effect of coarse pumice soils, volcanic eruptions, fires and a grim climate combined to 'stop the clock' – keeping the ecosystem in a pioneer stage and slowing the development of forests. Yet around the edges of the plateau the great forests of the central North Island – Mamaku, Whirinaki and Pureora – survived and thrived. Today, through the efforts of science and technology, the volcanic soils support the country's most important plantation forests, based on fast-growing North American conifers like *Pinus radiata*.

Why have podocarps continued to dominate much of the forest at Whirinaki and Pureora when the successional trend would have been towards lower-stature forest dominated by hardwoods such as Tawa, Kamahi and Rewarewa? One theory is that soil development was slowed through the rejuvenating effect of volcanic ash added by successive eruptions. Although we know that the AD186 Taupo eruption devastated the vegetation on the adjacent Kaingaroa Plateau, it is possible that parts of Whirinaki and Pureora escaped the full incinerating force of the pyroclastic flow, leaving enough seed trees and birds to maintain a podocarp forest.

The plant cover in Tongariro National Park and the southern Kaimanawa Range also poses many questions, for it is a sometimes puzzling jigsaw of tussocklands, shrublands, forests and harsh alpine gravel-fields. In Tongariro National Park there are

The Lady Knox Geyser in the Waiotapu geothermal field is a natural boiling spring artificially modified to provide a tourist attraction: soap is added so that the geyser erupts on time each day.

pronounced climatic differences between the higher alpine regions and the less severe margins of the surrounding plain, and between the more forested (and wetter) western slopes and the open tussocklands, herb-fields and windswept gravel-fields of the so-called Rangipo Desert in the east. But it is the complex history of eruptions that accounts for many of the ecological enigmas – such as the survival of dense podocarp forests only on the predominantly south-west slopes, which provided shelter from the devastating blasts of the Taupo eruption 1,800 years ago. A further curiosity is the inability of the normally tenacious beech forests to have recolonized the Tongariro massif.

The other puzzling locality is tucked away at the head of the Rangitikei River, between the southern end of the Kaimanawa Range and the north-west corner of the Ruahine Range. This Moawhango country is quite unlike any other part of the North Island, being extensively covered in Red Tussock and 'islands' of upland conifer forest (mainly Kaikawaka, with Hall's Totara, Mountain Toatoa and Pink Pine). It includes the oldest landform in the North Island – a remnant greywacke peneplain from the Cretaceous period, of similar age to the ancient peneplain of Central Otago (see page 169). Indeed, the area's landform and floristic similarities with Central Otago, 1,000 kilometres (620 miles) to the south-west in the southern part of the South Island, are quite striking, with both localities having a very diverse indigenous flora: 750 species for Moawhango and 910 for the much larger Central Otago, which also has wider variations in terrain and climate.

Urewera National Park, far to the east, has been less affected by volcanic eruptions and survives as the largest forested wilderness in the North Island. The local Maori people, the Tuhoe, refer to this vast complex of forest as Te Urewera. The Tuhoe are known as the 'Children of the Mist', in reference to their traditional belief that they are the offspring of Hine-pukohu-rangi, the celestial Mist Maiden. But their name aptly captures the almost ethereal qualities of the mist-shrouded ranges and valleys. This wilderness has probably sustained Maori occupants for more than 1,000 years, and was one of the last rebel strongholds in the land wars of the 1860s and 1870s. There is much that is tragic about the history of the inhabitants of Te Urewera; they are an integral part of the park, and some of their story is graphically told in the visitor centres at Aniwaniwa and Taneatua.

The immensity of Te Urewera is daunting to the first-time visitor. The topography is extremely rugged, the podocarp, hardwood and beech forests are wet and gloomy, and there are virtually no open tussock tops above the bush-line, all the ridge-crests seeming to unite into a jumbled forest maze. So many of the natural secrets of this forest wilderness still wait to be unravelled. The geology is very interesting, with huge earthquake-induced landslides forming Lake Waikaremoana as well as the fascinating 'flow landscape' in the softer mudstones and sandstones around Lake Waikareiti. It is here that the mis-named 'tundra' wetlands occur, a colourful mosaic of shallow lakes at about 900 metres (2,950 feet). These wetlands are surrounded by Red and Silver Beech forest on the drier slopes, but their margins contain a variety of subalpine plants more commonly found above the tree-line.

By comparison with New Zealand's coastlands and lowlands, the volcanic lands are not regarded as a region of high biological diversity. However, the forests are important habitats for a number of birds which are now severely restricted in distribution – especially the Kokako, Brown Kiwi, Kaka and Blue Duck. There are also some unusual plants, such as the ferns and fern relatives with tropical affinities growing in the warm 'steam zone' of the geothermal areas, and the brilliantly coloured red and blue-green algae which survive in the hot mineral waters. Another fascinating plant found only in this region (and on Little Barrier Island) is the Wood Rose *Dactylanthus taylori*. This rare forest-floor dweller is the only New Zealand member of the tropical Balanophoraceae family, an unusual group of non-photosynthetic plants which depend entirely on a host plant for their nutrients. Recent research has shown how the Wood Rose has a close relationship (in its pollination) with another extremely rare species: the Short-tailed Bat, the rarer of the two species of bat that are New Zealand's only indigenous land mammals.

Tongariro National Park

There are several reasons why Tongariro National Park is New Zealand's best known National Park: the unpredictable and violent behaviour of its volcanoes, its dramatic setting in isolation on the 'roof' of the North Island, and its popularity as the main winter skiing area for most of New Zealand's population.

The park, of area 76,500 hectares (295 square miles) is one of New Zealand's two World Heritage sites, and is considered by UNESCO to be of 'outstanding universal value'. In fact, it has been put on the World Heritage list twice, first in 1990 for its natural properties (its uniquely active composite volcanoes) and again in 1993 for its significance as a cultural landscape – for Tongariro has a strong historical connection with the Maori people of the Ngati Tuwharetoa tribe, through their origins in the Hawaikis of the Pacific to the Arawa canoe and Ngatoroirangi. But Tongariro is more than a mountain. It is spoken of by the Tuwharetoa with the same reverence they accord to an ancestor, for their identity is synonymous with this volcanic landscape:

Ko Tongariro te maunga.
Ko Taupo to moana.
Ko Ngati Tuwharetoa te iwi.
Ko Te Heuheu te tangata.

Tongariro is the mountain.
Taupo is the lake.
Ngati Tuwharetoa are the people.
Te Heuheu is the man [paramount chief].

To both Maori and Pakeha, Tongariro is cherished as the gift from Horonuku Te Heu Heu Tukino to all the people of New Zealand, and as such it became on September 23, 1887, New Zealand's first National Park, the world's fourth and the first freely given by an indigenous people.

ABOVE The small native owl – the Morepork or Ruru (*Ninox novaeseelandiae*) – is often heard at night mournfully calling its name, 'more-pork'. It is found in pine forests and parks as well as lowland podocarp/broadleaf forest. It eats insects, lizards, birds, rats and mice.

LEFT The lava-slopes of Mount Ruapehu are generally snow-covered in winter. Alpine plants often shelter in the lee of the andesitic lava boulders.

RIGHT Sand and scoria accumulate around boulders in the lava-fields allowing characteristic subalpine plants to establish: mats of Silver Raoulia (*Raoulia albosericea*), Bristle Tussock (*Rytidosperma setifolia*), Mountain Inaka (*Dracophyllum recurvum*) and Snow Totara (*Podocarpus nivalis*).

ABOVE Seen across Lake Rotoaira are the northern slopes, around Ketetahi, of the squat massif of Tongariro. This sprawling landscape has at least 11 volcanic vents and craters.

BELOW Alpine herbs like the mauve-flowered *Parahebe hookeriana* and the white-flowered Everlasting Daisy (*Anaphalis bellidioides*) abound along the Mangatepopo Stream in Tongariro National Park.

RIGHT The Rangipo 'Desert' to the east of Mounts Ruapehu and Ngauruhoe in fact receives about 1,200 millimetres (47¼ inches) of precipitation annually. Hardy low-growing *Pimelea, Carmichaelia, Gentiana* species and Bristle Tussock withstand the scouring winds and the extremes of summer heat and winter cold on this subalpine gravel-field.

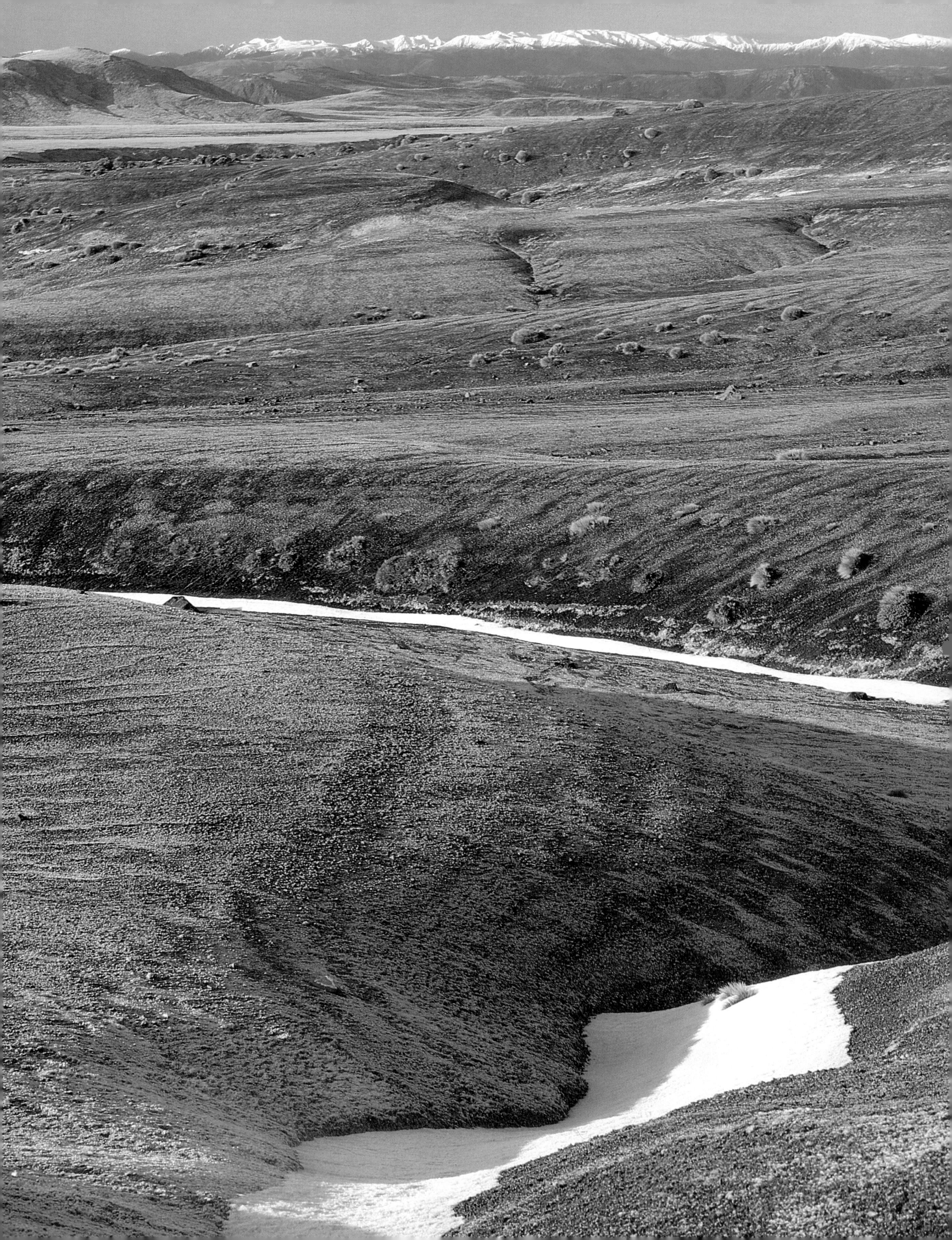

LEFT Looking south to Mount Ruapehu – 2,797 metres (9,177 feet) – from above the eastern lip of Blue Lake Crater on Mount Tongariro. Mount Ngauruhoe is the symmetrical volcano in the centre.

BELOW The Tama Lakes occupy explosion craters formed about 10,000 years ago in the broad saddle between Mounts Ruapehu and Ngauruhoe.

ABOVE The Whakapapa, Mangaturuturu, Manganui-a-te-ao and Mangawhero headwaters of the Whanganui River rise on the western slopes of the volcanoes of Tongariro National Park.

BELOW The Ketetahi Hot Springs, at 1,400 metres (4,560 feet) on Mount Tongariro's northern slopes.

Lake Taupo

Most New Zealanders and overseas visitors do not think of Lake Taupo as being part of a volcano; instead, this high – 357 metres (1,170 feet) above sea level – and deep – up to 160 metres (525 feet) – cold, clear lake is famous for its scenery and the quality of its fishing (for introduced rainbow trout). Yet it is a caldera lake occupying the site of the most violent eruptive vent in the Taupo Volcanic Zone. The caldera is enormous: with an area of 606 square kilometres (234 square miles), this is the largest of New Zealand's many lakes.

Today visitors have to look carefully for the signs of the great AD186 Taupo eruption. They are subtle, but everywhere: in the White Cliffs beneath the finely dissected Ouaha Ridge on the north-eastern shores of the lake; in the great depth of Taupo Pumice in the road cuttings on State Highway 5 just above Taupo township; or in the prominent beach remnants north of Hatepe, marking the level to which the new lake rose – over 30 metres (100 feet) above its present level – when the outlet was blocked after the great eruption. Lake Taupo is quiet now, but certainly it will erupt again – although we do not know when.

To the north of Taupo township and along the banks of the Waikato River, eerie steam clouds rise from fumaroles alongside State Highway 1. Some of this geothermal energy has been harnessed for electricity and other industrial purposes, destroying the once-famous Spa and Wairakei geyser-fields. One formerly minor geothermal field, Craters of the Moon (Karapiti), has by contrast increased its activity following the exploitation of the adjacent Wairakei Valley. Visitors to Craters of the Moon can freely wander a boardwalk among a profusion of erupting mudpools and fumaroles with colourful lichen, moss, Umbrella Fern and Manuka vegetation. Close by, in another awesome display of nature's power, the waters leaving Lake Taupo have cut a narrow chute in the ignimbrite walls to plunge out as the spectacular Huka Falls.

Lake Rotoaira, between Mount Tongariro and Tongariro National Park's three northern volcanoes, Pihanga, Kakaramea and Tihia. Lake Taupo is in the distance.

ABOVE Across Lake Taupo is the dormant volcano of Mount Tauhara behind Taupo township. Lake Taupo, New Zealand's largest lake, occupies the huge caldera of Taupo volcano.

BELOW LEFT Volcanic plugs, common around Lake Taupo, are often pieces of more resistant, welded volcanic rock left after the surrounding tephra has been eroded away.

BELOW RIGHT The 22-metre (72-foot) Huka Falls, about six kilometres (3¾ miles) downstream of Lake Taupo, where the Waikato River cannot cut down through a band of silicified rock.

LEFT The Wairakei geothermal field, near the Waikato River about seven kilometres (4½ miles) downstream of its outlet from Lake Taupo, is the largest geothermal field in New Zealand, and in places is exploited for electricity.

BELOW LEFT Not all the geothermal displays were lost when the Wairakei geysers and fumaroles were harnessed for electricity. The nearby Karapiti thermal area has steadily increased in activity, and is now an impressive geothermal attraction. Because the whole system is changing, with fresh fumaroles appearing regularly, the site has earned a new popular name – Craters of the Moon.

BELOW The vegetation of the Craters of the Moon geothermal area is subjected to localized heat and wafting steam. A colourful, low, semi-prostrate cover of *Lycopodium*, Umbrella Fern (*Gleichenia* spp) and stunted Manuka is found throughout much of the valley.

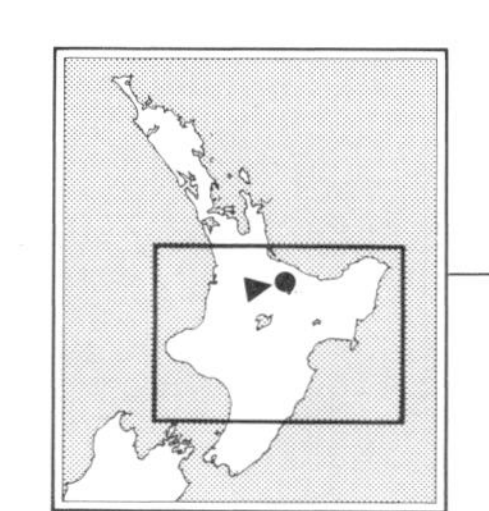

Rotorua

Rotorua is one of the geothermal wonderlands of the world. Its boiling pools, geysers, fumaroles and presentations of Maori culture have given the area an enduring attraction for the international tourist. While some of the trappings of this mass tourism can be criticized, there is still a wildness and unpredictability about Rotorua's ever-changing volcanic landscape that fascinates residents and visitors alike.

Lakes feature in the landscape every bit as much as volcanoes. Rotorua is the 'Lake District' of the North Island, with 11 major lakes lying within a roughly triangular region only 650 square kilometres (250 square miles) in area – not much larger than Lake Taupo. Unlike the great glacial lakes of the South Island, the Rotorua lakes all occupy calderas and other depressions caused by volcanic eruptions. Rotorua city is unique in its precarious location, not only on the edge of a volcanic lake (Lake Rotorua) within a superbly preserved circular caldera but also at the foot of a large rhyolite dome (Ngongotaha). Furthermore, it straddles two very active geothermal fields, at the Maori settlements of Ohinemutu and Whakarewarewa. The latter, world-famous for its Pohutu Geyser and other geothermal features, is New Zealand's last remaining unexploited geyser-field.

Out of sight beyond the caldera rim, but only 25 kilometres (15.5 miles) away to the south-east, lies Mount Tarawera. The huge line of chasms across its summit, Lake Rotomahana and the steaming cliffs at the bottom of the Waimangu Valley are products of the destructive Tarawera eruption of 1886. Visits to Waimangu and the other three main geothermal attractions around Rotorua – Whakarewarewa, Tikitere and Waiotapu – can be unescorted or carried out in the company of experienced guides. A visit to Whakarewarewa can be combined with the cultural experience of visiting the Maori Arts and Crafts Centre.

The Lake Tarawera landscape was severely affected by the 1886 eruption of Mount Tarawera, half-hidden by cloud on the lake's eastern side.

ABOVE Pohutu, the largest geyser in Rotorua's Whakarewarewa geothermal field, erupts about 15 times a day, the impressive 18-metre (60-foot) column creating a sinter mound several metres high.

ABOVE The colourful Champagne Pool, the largest hot spring in the Waiotapu geothermal field, is a 900-year-old explosion crater now filled with alkaline water.

BELOW Beneath Inferno Crater, the stream draining the Waimangu geothermal field contains several continuously active small hot springs coloured by green and brown algae, and possibly arsenic and tungsten minerals.

BELOW Carbon-dioxide-rich water from Champagne Pool cascades over the sinter rim, giving off CO_2 gas and depositing silicates and other minerals.

ABOVE *Baeomyces*, New Zealand's commonest soil-inhabiting lichens. Pink, spore-producing fruiting bodies extend from the white thallus, which appears as a film on the soil.

RIGHT Warbrick Terrace, in the Waimangu field near Lake Rotomahana, owes its spectacular colour to algae and to iron oxides and hydroxides mixed with the silicate deposits.

BELOW LEFT Whakarewarewa, now the only major unexploited geyser-field in New Zealand, is famous for its more than 500 hot springs, many now enclosed in their own sinter pool.

BELOW RIGHT The 'Emerald Pool' occupies the southernmost Waimangu Valley crater formed in the 1886 Tarawera eruption. This cold lake is coloured by sphagnum moss in its shallows.

The main 1886 rift across the summit of Mount Tarawera, among the most desolate yet colourful places in central North Island's Volcanic Plateau. The dark rocks are basaltic scoria from the 1886 eruption; the grey and white rocks are pumice and rhyolitic lava from earlier ones.

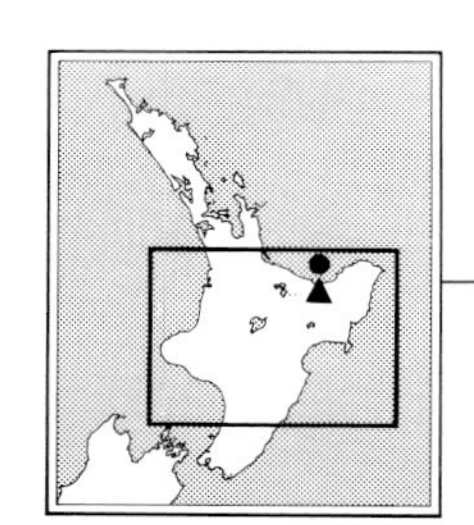

White Island

White Island, or Whakaari, lies 50 kilometres (30 miles) offshore in the Bay of Plenty, its ever-present plume of white steam indicating a state of constant volcanic activity – in fact, White Island is the most active volcano in New Zealand and the northernmost and most inaccessible in the Taupo Volcanic Zone (although there are active submarine volcanoes between White Island and the Kermadecs). Because of the difficulties of access it has only recently started to become a significant tourist attraction, although for some years earlier this century it was mined for sulphur, until the combination of loss of life, operational hazards and poor returns made the venture unsustainable. White Island is unusual by comparison with the rest of the North Island's active volcanoes in that its vent, although below sea level, is sealed from contact with sea water. The violent geothermal activity is the result of rainwater seeping down into contact with a very shallow magma source.

Despite its stark, arid-looking and beautifully wild landscape, the island supports a very select band of hardy natural colonists. The ubiquitous coastal tree Pohutukawa forms a virtual monoculture in a few sheltered locations, although none of the trees are more than 180 years old. The few shrubs and herbs which survive are able to withstand the combined hostile effects of salt-laden winds and corrosive, acid-producing gases.

There are three large breeding colonies of gannets on the southern headlands. With numbers stable at about 5-6000 pairs, White Island is as important a site for the Australasian Gannet as Hawke's Bay's spectacular Cape Kidnappers. The marine environment around the island is almost as diverse as that of the Poor Knights Islands (see page 64); Kingfish and other big gamefish attract deep-sea anglers, and for divers there is the added excitement of submarine volcanic vents.

White Island (Whakaari), the most active volcano in New Zealand, lies 48 kilometres (30 miles) offshore in the eastern Bay of Plenty.

ABOVE White Island fumaroles have a toxic mix of steam, sulphuric acid and ash. Over 10,000 tonnes of sulphur were mined here before 1934.

BELOW LEFT White Island's volcanic landscape has been active at least 15,000 years. The small vents within the outer crater walls have changed markedly since records began.

BELOW The spectacular geothermal activity arises as rainwaters mix with magma at shallow depths in the throat of a 700-metre (2,300-foot) volcano rising from the seafloor.

Despite the constant volcanic activity on White Island, 5-6,000 Australasian Gannets breed on the headlands at the southern end.

They plummet vertically into the clear waters to feed on the rich marine life around the island.

Whirinaki Forest Park

Sandwiched between the huge ignimbrite plateau of the Kaingaroa Plains and the steep greywacke slopes of the Huiarau Range lies the remote Whirinaki Valley, with one of the most remarkable rainforests in the world. This timeless place is the traditional home of the Ngati Whare people, and a forest stronghold from which the Maori prophet and guerilla leader Te Kooti harassed the government forces in the late 1860s. Just over 100 years later, Whirinaki was the scene of one of the most bitter forest conservation battles in the North Island – the decade-long campaign to save this priceless remnant of the great dense podocarp forests that once covered the lowlands and volcanic basins.

Whirinaki is not easy to penetrate. Although it is now managed as a forest park, 55,000 hectares (212 square miles) in area, it is far from the main tourist routes and has no prominent landmarks, and the network of logging roads serving the plantation forests around its periphery is confusing. Good information can be obtained from the Department of Conservation's Te Ikawhenua Visitor Centre, just beyond Murupara on State Highway 38, which leads to the fastness of Te Urewera. But, sadly, the world now seems to have abandoned the forest that in the early 1980s was a household name synonymous with the struggle for conservation.

What so impresses any visitor who perseveres to penetrate this sombre forest is the sheer size and density of the podocarp trees. Sir Charles Fleming, the famous scientist and naturalist, used the term 'podocarp gothic' to evoke comparisons with the great cathedrals of Europe, and truly such comparisons are there to be made. All five main podocarp trees – Rimu, Miro, Matai, Totara and Kahikatea – grow in profusion, sometimes together, sometimes in purer stands. Great trunks support a vaulted canopy of arching branches. Their height is daunting – up to 65 metres (213 feet) for Kahikatea – and often the top of the tree cannot be seen from inside the forest.

Whirinaki Forest Park is the outstanding dense podocarp and podocarp/broadleaf rainforest wilderness of the North Island.

ABOVE Where the Whirinaki River squeezes through the narrow Te Whaiti-nui-a-toi Canyon, all manner of ferns reach out from the canyon walls. Overhead, the crowns of densely packed Kahikatea, Rimu, Matai and Totara trees filter the light entering this primaeval place.

BELOW Rata vines are characteristic of the Whirinaki and Urewera forests. Many Rata species start life as an epiphyte high in the limbs of an old podocarp tree; then, gradually, their roots enclose the host's trunk.

ABOVE The outstanding impression on visiting the forest wilderness of Whirinaki is of the sheer size and density of the trunks of the magnificent Rimu, Kahikatea and Totara trees.

BELOW In their search for forest foods, Kaka (*Nestor meridionalis*) help distribute pollen and seed. The bird's future is threatened by introduced wasps and destruction of forest habitats.

Pureora Forest Park

Only one other dense podocarp rainforest comes close to matching Whirinaki in magnificence: Pureora Forest Park. Pureora likewise thrives in a volcanic landscape of ignimbrite overlain by airfall ash, and the same species of tall podocarp trees stand out above a canopy of Tawa, Kamahi and Rewarewa. This region was in 1978 the scene of another high-profile conservation campaign – which included protesters remaining for weeks in the crowns of trees about to be felled. The crusade captured national attention and exposed the unsustainable logging of the remaining old-growth forest. As a result, a large proportion of the main forest types were strictly protected in ten ecological areas extending 60 kilometres (37 miles) along the crest and slopes of the Rangitoto and Hauhungaroa ranges, which form the western watershed to Lake Taupo.

Pureora is a little closer to the main tourist routes than Whirinaki. For the road traveller it provides an easily approached forest interlude during the journey from the Waitomo Caves to Rotorua; the short diversion off State Highway 30 to the park information centre is amply rewarded. Two of the most interesting ecological areas are within easy access. The first, Pikiariki, features the Totara Walk (with superb specimens of the main podocarp trees and an accompanying interpretative booklet on the forest), a forest observation platform and the remarkable 'buried forest' (relics of the AD186 Taupo eruption preserved in a bog until they were accidentally discovered – by a bulldozer! – in 1983). The other ecological area worth visiting is Pureora Mountain. A track to the summit of this old volcano passes through the best altitudinal sequence of forest in the park.

Pureora Forest Park straddles the Hauhungaroa Range west of Lake Taupo. Kahikatea, Miro, Rimu and Totara soar above a canopy of Tawa, Kamahi and Rewarewa.

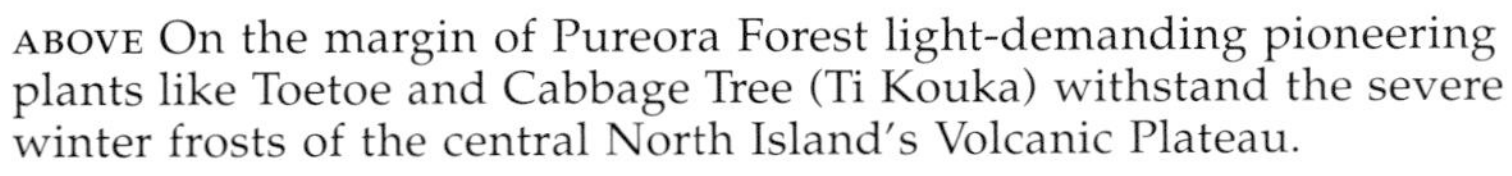

ABOVE On the margin of Pureora Forest light-demanding pioneering plants like Toetoe and Cabbage Tree (Ti Kouka) withstand the severe winter frosts of the central North Island's Volcanic Plateau.

ABOVE RIGHT The tree fern *Dicksonia fibrosa* is common in Pureora Forest, especially in cold sites and gullies. From above it shows amid the leafy canopy as a bright green star-like motif.

RIGHT Foliose *Stricta* lichens are common in moist forest, usually growing on tree trunks in association with mosses.

BELOW LEFT Young Bellbirds (*Anthornis melanura*), given to mimicry, learn their beautiful bell-like song from neighbouring adults. They eat insects, fruit and nectar, pollinating many plants and spreading seeds.

BELOW RIGHT Restricted to several hundred hectares of gorse-covered farmland near Pureora Forest, the Mahoenui Giant Weta (*Deinacrida* spp) is defended from predators by the harsh spines of the gorse – ironically, an introduced weed.

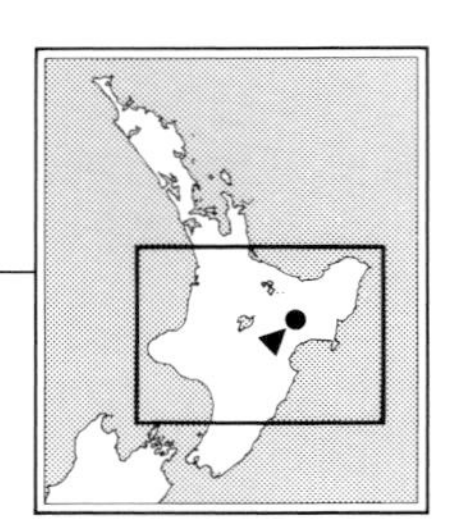

Urewera National Park

Urewera National Park is the largest forested wilderness remaining in the North Island. It lies high and remote, straddling the Huiarau Range, part of the great axial greywacke spine, 700 kilometres (435 miles) long, of the North Island. At 213,000 hectares (820 square miles) it is the fourth largest National Park in New Zealand and the largest by far of those in the North Island.

Most visitors reach the park via the tortuous State Highway 38, the main focal point being Lake Waikaremoana with its motor camp, Aniwaniwa Visitor Centre and the network of short walks. There is also a major three- to four-day walking track (the Lake Waikaremoana Track) around the indented southern and western shores of the lake. The lake is high – 580 metres (1,900 feet) above sea level – cold and popular for boating and trout fishing.

The Huiarau Range was one of the last refuges of a beautiful, now extinct wattlebird, the Huia. However, a large population (nearly 300 birds) of its blue-wattled relative, the Kokako, was 'rediscovered' as recently as 1991 in the Waimana Valley. The discovery gives hope that the very rugged northern Urewera may be the largest stronghold of the bird in the North Island.

One of the walks showing the range of forest vegetation in the park is a gentle hour-long climb through Northern Rata-Rimu-Tawa forest around the shores of Lake Waikaremoana, to the Red and Silver Beech forest fringing Lake Waikareiti, which lies at an altitude of 880 metres (2,885 feet). A wide range of forest birds can be seen and heard in Lake Waikaremoana's surrounds; Brown Kiwi (found throughout the park), Kereru (New Zealand Pigeon), Kaka, parakeets, Whitehead, Paradise Shelduck and the inquisitive North Island Robin are some of the more visible or vocal. The fast-flowing rivers draining the park are the usual habitat for another distinctive bird, the whistling-duck called Blue Duck or Whio.

Lake Waikaremoana ('Sea of Rippling Waters') lies 580 metres (1,900 feet) above sea level in the ranges of Urewera National Park – Te Urewera.

ABOVE Nearly 300 metres (985 feet) above Lake Waikaremoana, Lake Waikareiti ('Little Rippling Waters') is studded with islands and fringed with beech forests. This beautiful lake, reached by walking track from the Aniwaniwa Visitor Centre, has a unique charm, being completely unmodified by introduced lake weeds. Power boats are banned.

ABOVE RIGHT These huge blocks of sandstone on the slopes above Rosie Bay are part of a massive landslide off the Ngamoko Range about 2,200 years ago. The debris blocked the Waikaretaheke River and water filled the valley to a depth of 240 metres (790 feet), forming Lake Waikaremoana.

RIGHT Many of the rivers draining into Lake Waikaremoana plunge over wide sills of sandstone close to the lake's edge. The Aniwaniwa River has a series of these waterfalls, the most spectacular being Papakorito, about two kilometres (1¼ miles) upstream from the park's visitor centre.

ABOVE The New Zealand Scaup is a bird of still waters, living mostly in deep lakes like Waikareiti. Once hunted heavily, it is now totally protected and making a good recovery.

BELOW Paradise Shelduck are often heard before seen, the male honking loudly if disturbed. They are common around the shores of Lake Waikaremoana and in the park's river valleys.

ABOVE The Kokako is related to other wattlebirds like the Saddleback and the extinct Huia. Urewera National Park's forests are among its last strongholds.

FAR LEFT Kakabeak (*Clianthus puniceus*) is easily cultivated but rare in the wild owing to habitat loss and popularity with introduced browsers. The few plants left are mostly around Urewera National Park.

LEFT The pure white Clematis (*Clematis paniculata*) flowers are startling against the dark foliage of the forest. The plant's leaf stalks twine around slender supports up to the canopy.

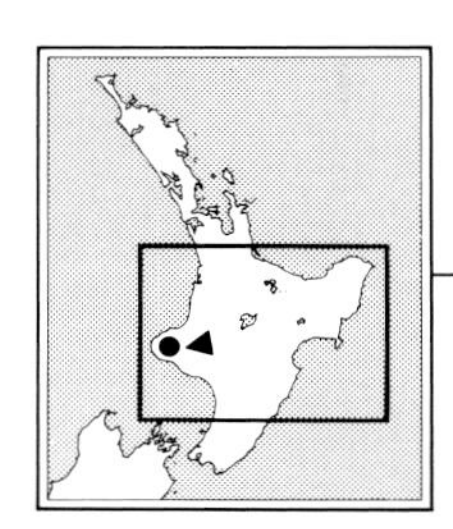

Egmont National Park

Mount Taranaki (Mount Egmont) is to the North Island of New Zealand as Mount Fuji is to Japan. From every direction the beautifully symmetrical cone of this dormant andesitic volcano, 2,518 metres (8,256 feet) tall, dominates the skyline. It stands on its own huge ringplain in splendid isolation far to the west of the other volcanoes of the Taupo Volcanic Zone (see page 84). Egmont National Park is highly valued by the people of Taranaki, whose campaigns ensured its protection in 1900 as New Zealand's second National Park. Of area 33,540 hectares (130 square miles), the park protects all the forest, herb-fields and lava slopes above about the 400-metre (1,310-foot) contour on the three cones of Taranaki, Pouakai and Kaitake.

Because of its height and proximity to the coast, Mount Taranaki is subject to very rapid changes in weather; these, coupled with steep slopes that are generally icy in winter, have earned it the notorious reputation of being the most dangerous of New Zealand's more accessible mountains. The high rainfall – up to 8,000 millimetres (315 inches) annually at the tree-line – and milder coastal climate have given the rainforest a verdant character that is matched elsewhere only in the Tararua Range, north of Wellington, and on the West Coast of the South Island. The lower forest slopes have scattered Rimu over a tight canopy of Kamahi, Fuchsia and a variety of broadleaved trees, their branches heavily festooned with trailing strands of moss. Huge isolated Rimu and Northern Rata trees soar above this canopy. At higher altitudes, Mountain Totara and Kaikawaka (a striking-looking mountain cedar) dominate, to be replaced at the tree-line by an almost impenetrable shrubland aptly dubbed 'leatherwood' by trampers from the lower North Island.

There are many plant enigmas attached to Egmont National Park, including the complete absence of beech forest. In this volcanic 'island' alpine plants have evolved in isolation from those of other alpine regions of the North Island and north-western South Island.

Pararaki ('Seagull Rock') and Mataora ('Round Rock'), two of the Sugarloaf Islands, are fragments of Taranaki's oldest volcano.

ABOVE Tongariro National Park's volcanoes seen by dawn light from the slopes of Mount Taranaki (Egmont). Tongariro and Ngauruhoe are on the left, the massif of Ruapehu on the right. The basin of the Whanganui River lies under the cloud in the centre.

LEFT Mount Taranaki's middle slopes are covered with dense montane rainforest. The tight-knit canopy is of Kamahi, Fuchsia, Hall's Totara and Kaikawaka. The interior is sometimes called 'goblin forest' because of the gnarled shape of the trees and the trailing moss hanging from the boughs.

BELOW LEFT Karamu (*Coprosma lucida*), found in drier areas within regenerating forest throughout the country, is among the easiest recognized of the larger-leaved Coprosmas. Early European settlers roasted and ground the berries as a coffee substitute.

BELOW Large *Pseudocyphellaria* lichens are conspicuous in the Egmont National Park's forest. The leaflike form grows slowly but abundantly on branches and tree-trunks, its yellow underside mainly noticeable when the lichen is dislodged onto the forest floor.

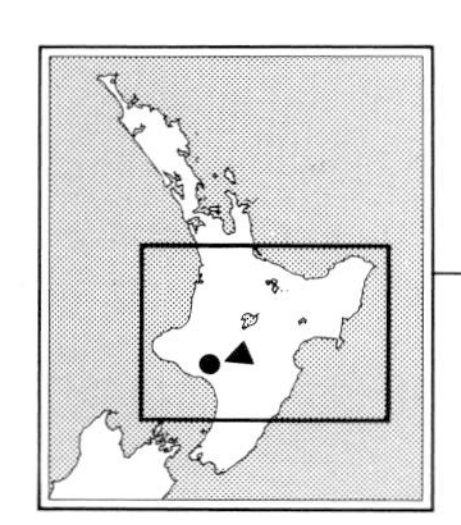

Whanganui National Park

Whanganui National Park is not only one of New Zealand's newest National Parks, it is also geologically one of the youngest. The landscape is a maze of intricate ridges and V-shaped valleys eroded from the local soft mudstone (commonly referred to by its Maori name *papa*). Most of the streams and rivers eventually flow into the park's main artery, the Whanganui River, which originates in the volcanic lands high on the slopes of Mount Ruapehu. For 170 kilometres (105 miles) the sinuous Whanganui winds its way through the park, providing a significant attraction for canoeists and jet-boaters because of its tranquillity, forest scenery and deep historical and cultural significance to Maori and Pakeha alike.

The headwaters of the Whanganui are clear, as with the rest of the rivers on the Volcanic Plateau. However, they quickly become turbid with suspended silt- and clay-sized sediments as they cross the geological boundary from pumice to *papa*. The clearer waters of the Whanganui's eastern tributaries (for example, the Whakapapa, Retaruke and Manganui-a-te-ao) are among the most important remaining North Island habitats for one of the world's most interesting ducks, the Whio or Blue Duck, a torrent duck admirably suited to fast water. The Whio is endemic, of ancient origin and, like many of New Zealand's more unusual birds, declining in numbers.

The central part of the park is the largest area of lowland forest in the North Island. It is fairly uniform: a canopy of broadleaved trees like Kamahi, Tawa and Hinau is overtopped by large isolated trees of Rewarewa, Northern Rata and Rimu, plus other podocarps like Totara and Kahikatea. Distinctive features of the forest are the profusion of tree ferns in the gullies and the Black Beech trees hugging the dry sandstone ridges.

The Whanganui River, clear where it rises on Tongariro National Park's volcanoes, soon picks up silt and clay from the soft mudstone country.

ABOVE Blue Duck (*Hymenolaimus malacorhynchos*) can swim against the current, and chicks feed in fast-flowing rivers as soon as hatched. Few ducks can occupy this type of habitat.

LEFT The papa (mudstone) cliffs are a spectacular accompaniment to a trip on the Whanganui River. Rimu and Rata reach up above a canopy of Tawa and Kamahi, and ferns drape the shadier cliff faces.

BELOW Omorehu Waterfall, one of hundreds in Whanganui National Park. The papa rocks making up most of the park are easily eroded; where a harder sandstone lip persists, a waterfall develops.

COOK STRAIT, NELSON AND MARLBOROUGH

The crossing of Cook Strait by ferry, from Wellington to Picton, can often be a wild voyage. On a good day, however, it offers superb views of the bold uplifted marine terraces around Baring Head (to the east of the entrance to Wellington Harbour), the gale-lashed cliffs on both sides of the strait, and the sinuous channels of the Marlborough Sounds. But this is just the beginning, for the northern part of the South Island probably has the most varied landscape of any New Zealand region. This is surprising considering there are no volcanoes or glaciers and few lowland plains of any significance.

On arrival at Picton, three very different highly scenic road routes into Nelson and Marlborough offer the visitor a tantalizing prospect of diverse landscapes and wildlife observation. The 'Top of the South' heritage highway leads to Farewell Spit far off in the north-westernmost corner of Nelson. This route traverses a bewildering array of rock types – greywacke and schist, marble and serpentinite. Around Tasman Bay there are many attractive estuaries, including Nelson Haven and its unusual Boulder Bank. Further west, the road passes some of the best karst features in the country, especially the sculpted marble outcrops of Takaka Hill and Waikoropupu Springs in Golden Bay.

Another route heads into the mountainous interior, along the great fault-line followed faithfully by the Wairau River. It skirts the glaciated landscapes of Lake Rotoiti (a blue gem in its beech-forest setting) to plunge into the wild, forested, earthquake-prone gorges of the Buller River. Eventually, this route squeezes between the dramatic coastal limestone cliffs of Paparoa National Park and the Tasman Sea.

The third route – the eastern one through coastal Marlborough – is very different, but no less wild and scenic. Leaving the important estuary known as the Wairau Lagoons, it passes through the remnants of former low-altitude tussocklands. The contours of the hills are smooth and symmetrical, moulded by layers of loess (windblown silt) deposited long ago as the ice-age glaciers retreated from this barren landscape. The route then hugs the rocky Kaikoura coastline, with its pounding surf, Bull Kelp, seals and enticing glimpses of the harsh scree slopes of the Seaward Kaikoura Mountains towering above.

Picton is also the ideal town from which to undertake an exploration – by water – of the Marlborough Sounds. The coastline of the sounds has an intricacy unmatched elsewhere in the country – to the extent that this relatively small area contains 15 per cent of the national coastline. While so much of New Zealand is still being uplifted by tectonic (enormous geological) forces, the sounds offer a 'drowned landscape', where the northern tip of the South Island is being tilted into the sea. Although much of the original forest has now been removed, through largely fruitless attempts to farm the poor soils of the steep ridges, the scenic sheltered waterways of the inner sounds are very popular for boating, fishing and diving. There are many historical sites associated with the early European navigators, and tracks like the Queen Charlotte Walkway traverse the best of the forests, seascapes and historic sites.

The outer sounds, near the wild waters of Cook Strait, are isolated and windswept. Gale-force winds in excess of 160 kilometres (100 miles) per hour are regular, and have been the cause of many shipwrecks in what are probably the most turbulent waters around the main islands of New Zealand. The highest recorded wind gust in Cook Strait reached 267 kilometres (166 miles) per hour, on April 10, 1968; this was in the appalling storm that drove the inter-island ferry *Wahine* onto Barrett Reef at the entrance to Wellington Harbour. Yet in this least hospitable of coastal environments lie some of the most important wildlife-sanctuary islands in the country. Kapiti and Mana islands lie on the North Island side of Cook Strait; while in the outer sounds the key wildlife islands are the Brothers, Chetwodes and Trios, as well as Maud Island and an extraordinary wildlife sanctuary, Stephens Island (see page 120).

The geological and climatic diversity of the region is reflected in other seascapes besides the sounds. The limestone of the western coasts, constantly battered and scoured by the westerly winds and the Tasman Sea, has been sculpted into fantastic shapes – the 'Pancake Rocks' of Punakaiki, the stacks of the Paparoa coastline and the arches and cliffs around Whanganui Inlet, one of the most natural large estuaries left in the country. In contrast, granite is the source of the golden sands of the beautiful sheltered beaches of Golden Bay and Abel Tasman National Park. Limestone is again a feature on the east coast, at the Kaikoura Peninsula, where large colonies of New Zealand Fur Seal frequent impressive reefs and wave-cut platforms. Near Kaikoura the continental shelf is narrow, and the colder southern currents create a unique marine habitat frequented by Sperm Whales, Dusky Dolphins and the threatened Hector's Dolphin – all combining to provide New Zealand's premier marine ecotourism attraction. The need to conserve the best of this diverse regional coastal and marine environment has been recognized through the creation of marine reserves around Long Island in the sounds and Tonga Island on the Abel Tasman coast; there are also well advanced proposals for parts of Whanganui Inlet and Kaikoura Peninsula.

Everywhere throughout the region mountains press close to the sea. In the west, the Paparoa Range and Tasman Mountains are drenched in rain and often wreathed in fog. The Kaikoura Mountains in the east are dry and cold, often heavily mantled in winter snows from their exposure to southerly storms that race up the east coast of the South Island. The geological diversity of the region gives rise to mountains of widely differing shape and vegetation: the jagged spires of the hard gneiss of the Paparoa Range; the tussock-covered granite uplands of the Gouland and Gunner Downs in the Tasman Mountains; the remarkable Cobb Valley where the oldest fossils (trilobites) and volcanic rocks in New Zealand are found; the sparse serpentine vegetation of the ultramafic rocks of Dun Mountain and the Red Hills of the Mount Richmond Range; and the talus slopes and beech forests of the Spenser Range – to name just a few. An interesting feature of the mountain ranges is the way in which they are mostly oriented north-east/south-west, many of them in the east separated from

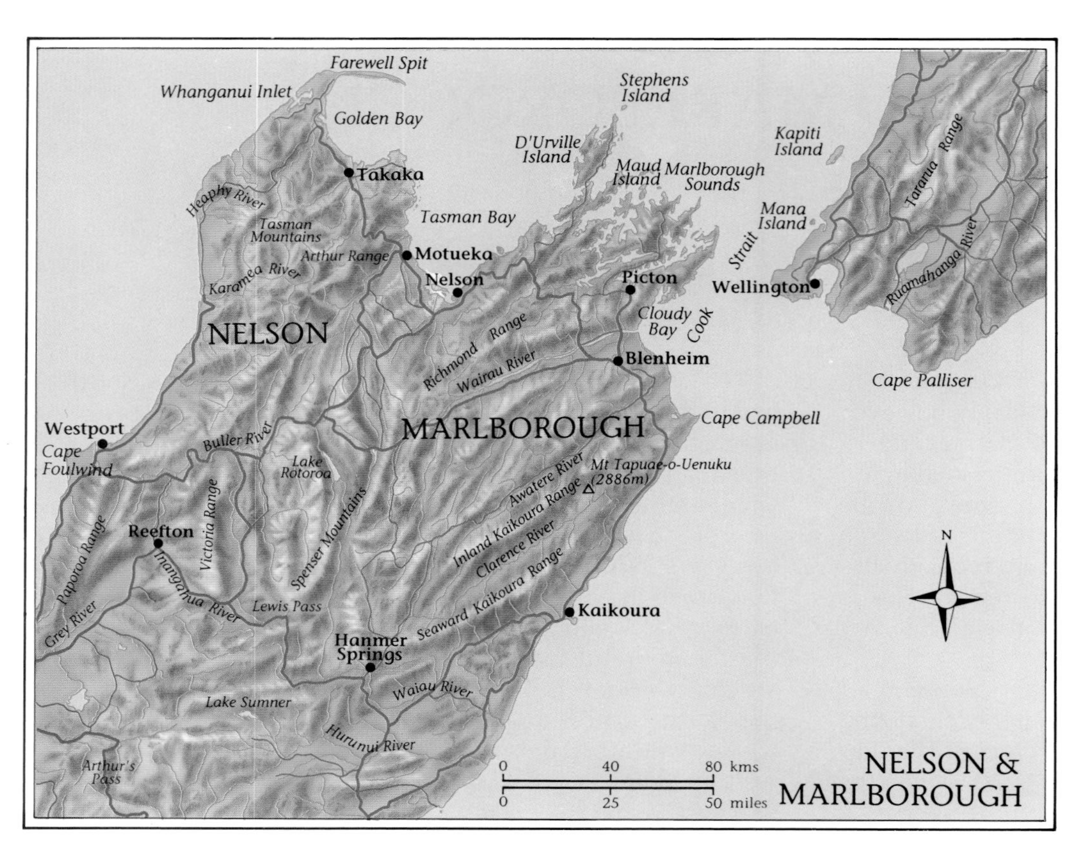

RIGHT D'Urville Island from Stephens Island, showing the bleak, windswept character of Cook Strait and the outer Marlborough Sounds. Gales regularly scourge Cook Strait at over 160 kilometres (100 miles) per hour and have caused many shipwrecks in probably the most turbulent waters around mainland New Zealand.

OPPOSITE BELOW Crossed by the Heaphy Track, the Gouland Downs, the largest of several Red Tussock (*Chionochloa rubra*) downlands in the heart of Kahurangi National Park's Tasman Mountains, are, with the surrounding montane beech forests, important habitats for Roa (Great Spotted Kiwi) and giant land-snails.

each other by very pronounced fault-lines. Here, in the north of the South Island, the great Alpine Fault splinters – and so do the Southern Alps, into a myriad lesser mountain ranges, running parallel to each other. Nelson and Marlborough may lack the alpine grandeur of Canterbury and Westland, but they more than make up for it in alpine diversity.

Because of the wide distribution of limestone and marble, the area contains New Zealand's outstanding karst landscapes. The lowland Punakaiki and Oparara limestone karst in the west are the most accessible, their river canyons and caves lying in luxuriant rainforest. The limestone arches over the Oparara River are considered the largest in Australasia. Nearby, the Honeycomb Hill caves contain sub-fossil bird remains (including those of an extinct goose and a giant eagle) dating back 20,000 years to the last ice age. But it is the wild and remote alpine karst in the marble mountains of Kahurangi National Park for which the area is famous. The glaciated alpine karst of Mount Owen is a landform of international significance; four of the cave systems of Mount Owen and Mount Arthur are, with depths of up to 900 metres (2,950 feet), the deepest in the Southern Hemisphere; and this alpine area has the three longest cave systems in the country, the most extensive reaching 34 kilometres (21 miles). In addition, the marble karst of the Pikikiruna Range in the uplands of Abel Tasman National Park is linked to the aquifer feeding the Waikoropupu Springs near Takaka, far below in Golden Bay. The springs are a major tourist attraction, and are protected as a scenic reserve; their cold waters are three times clearer than that of any other natural water body in the country, and this clarity and the spectacular array of freshwater flora attract an increasing number of divers.

Nelson and Marlborough are among the most significant areas of plant diversity in New Zealand. The mountains of north-west Nelson and south Marlborough are two of the five main centres of plant evolution and endemism on the mainland (the other three are Northland, Central Otago and Fiordland); this diversity has been generated by the wide variety of climates, landforms, geology and soils already discussed, with further important factors being

BELOW The view from the slopes of Mount Tapuae-o-Uenuku, the highest mountain in the Inland Kaikoura Range, across the Clarence Valley to the Seaward Kaikoura Range. The mountains of inland Marlborough differ greatly from the Southern Alps, being much drier, with warm dry summers and cold winters, and subject to southerly storms sweeping up the South Island's eastern coast.

the areas' isolation and historical stability: both areas were relatively undisturbed by the rejuvenating impact of the last glaciation.

The richness of montane and alpine plant-life in the vast mountains of north-west Nelson was a major consideration in its being made a National Park, New Zealand's most recent: Kahurangi National Park. The park contains half of the country's indigenous plants (1,226 of 2,450 species) and an extraordinary 80 per cent of New Zealand's alpine flora. Over 40 plant taxa are endemic to the mountains of the north-west, including several Forget-me-nots (*Myosotis* spp.), a *Clematis* adapted to soils on marble and limestone and, confined to either calcareous or ultramafic areas, shrubs of the ubiquitous genus *Coprosma*. Many trees, subalpine shrubs and alpine herbs reach their southern or northern distribution limit in the north-west. The sheer size and integrity of these uplands make the park a stronghold for many birds now eliminated from the lowlands: Great Spotted Kiwi, Kaka, Blue Duck (Whio) and New Zealand Falcon. Among the distinctive birds for which this area is the northern limit are the Kea, Rock Wren and Yellowhead. Kakapo probably survived in the Tasman Mountains until the last few decades.

Evidence for the uplands of the north-west acting as a plant and animal refuge during the Pleistocene ice age is particularly

A winter aerial view of Lake Tennyson (and its bordering moraines) and the eastern Spenser Mountains shows the largely treeless nature of the inland Marlborough landscape. The relatively high numbers of rare endemic plants in these drier inland basins make them a high conservation priority.

well illustrated by the fascinating range of indigenous invertebrates found in this remote corner of the South Island. There is a remarkable cave fauna, including at least 20 different cave wetas and nine cave beetles, most of which are endemic. Occupying this same habitat and preying on cave wetas is the Nelson Cave Spider; with its alarming-looking legspan of 12 centimetres (4.7 inches), this is easily New Zealand's largest spider. It is also one of the most ancient, belonging to the primitive Gradungulidae family, the last survivor of the various groups of spiders that flourished hundreds of millions of years ago.

The north-west uplands are also a centre of endemism for freshwater insects, like caddisflies and stoneflies. Here is the home of the rare Nelson Alpine Weta, one of the smallest of New Zealand's ten known species of giant weta, its stronghold being the tussock grasslands of the Mount Arthur Tablelands. But the colourful giant land-snails of the *Powelliphanta* genus, with their ancient lineage and curious distribution, are the true heralds of the uniqueness of this mountain environment. Nearly half of the 40 named taxa in the *Powelliphanta* genus, with diameters up to ten centimetres (four inches), are found in these uplands; all members of the oldest family of carnivorous land-snails in the world (the Rhytididae), tracing their origins back some 200 million years, they are mainly found in the cool, wet, montane beech forests or subalpine tussocklands, and may require the calcium-rich soils of the area for shell development. Some species are known to be long-lived, with lifespans of over 40 years.

South Marlborough, the other major centre of endemism, is very different from north-west Nelson. Here the mountains are higher and drier; they are much colder in winter and their geology is very different, with fault-shattered greywacke and argillite giving rise to vast areas of dry, open scree-field. Viewed across Cook Strait from Wellington, the two great mountain ranges – Inland Kaikoura and Seaward Kaikoura – make a majestic sight. The inland range is crowned by Mount Tapuae-o-Uenuku, at 2,886 metres (9,462 feet), and its companion peak, Mount Alarm, at 2,865 metres (9,393 feet), both much higher than the highest peaks of the North Island. The seaward mountains rise up sheer from the narrow Kaikoura coast to a height of 2,600 metres (8,525 feet). Far below, between the two rampart ranges, the wild and remote Clarence River twists and turns in search of a route through the formidable 100-kilometre (60-mile) mountain barricade to the Pacific Ocean.

To the south-west, tucked in between this barrier and the Spenser Mountains of the main divide, lie the sprawling intermontane basins of the Molesworth Country; Molesworth is New Zealand's largest cattle station. This is 'rain-shadow' habitat, remote and rarely visited, but containing a wealth of rare plants, including scree and rocky alpine herb-field communities.

Within this area of south Marlborough – and especially the Kaikoura ranges – there are numerous endemic giant wetas, skinks and geckos; it is the most diverse lizard habitat on the mainland. Despite 130 years of devastating pastoral fires and grazing, which have eliminated most of the forest habitats, the Seaward Kaikouras are still particularly interesting; they have at least three (possibly four) species of giant weta and the only eastern Kea population, plus the rare Weeping Broom communities and a most unusual marine-montane ecological relationship: here are found the only mainland breeding colonies of Hutton's Shearwater, high among the Snow Tussocks about 1,500-1,800 metres (4,900-5,900 feet) above sea level.

FAR LEFT *Powelliphanta hochstetteri obscura*, almost ten centimetres (four inches) in diameter and found at altitudes above 300 metres (984 feet), is among the very rich fauna of colourful giant land-snails in the mountains of Kahurangi National Park.

LEFT New Zealand's biggest spider, the insectivorous Nelson Cave Spider (*Spelungula cavernicola*), is a late survivor of a type that prospered millions of years ago. It is found in only a few cave systems, like the Oparara Caves in Kahurangi National Park and the caves of Paparoa National Park.

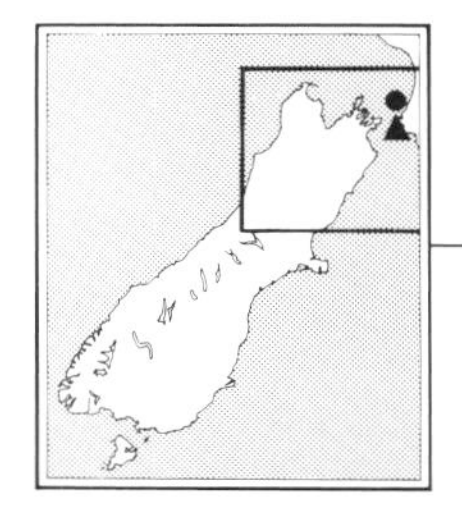

Kapiti Island

Kapiti Island lies six kilometres (3.7 miles) off the south-west coast of the North Island, about 50 kilometres (30 miles) north of the capital, Wellington. It is one of New Zealand's most important nature reserves, the largest in the wild Cook Strait area, and the closest to a major urban population, with up to 4,000 visitors per year permitted to disembark under carefully controlled conditions to learn about the birdlife. To many the island is also associated with the bloody exploits of Te Rauparaha, the fierce and cunning Ngati Toa chief who captured the island from Horowhenua tribes in the early 1820s. Later on there were attempts at farming, which destroyed about 75 per cent of the coastal broadleaf forest, and at whaling, which failed. The island's potential as a bird sanctuary was eventually recognized in 1897.

Today's visitors can have little idea of the scale of past environmental destruction, for the island is now almost completely covered in a regenerated forest of Tawa, Kohekohe, Northern Rata and Kanuka. Like Little Barrier Island (see page 75), Kapiti was the scene of a hallmark eradication campaign, representing the first time the tenacious Australian Brush-tailed Possum was successfully cleared from such a large island – nearly 2,000 hectares (7.7 square miles). It took six years of intensive tracking, hunting and poisoning to kill all 22,000 animals in this wild landscape of cliffs and gullies.

The most visible forest birds are Tui, North Island Kaka, pigeon, North Island Robin and North Island Saddleback. Less obvious are Stitchbird, Kokako and the Little Spotted Kiwi – a secretive nocturnal bird which owes its continued existence to its introduction to Kapiti, for it is now almost extinct on the main islands. Takahe have been introduced and seem to thrive in the shrubby forest margins, although Norway Rats and Kiore still pose a threat to the reintroduction of any further ground-nesting forest birds. The most recent conservation initiative has been to extend the zone of strict nature protection seawards, through the formation of the Kapiti Marine Reserve around part of the island and across to the Waikanae River estuary on the North Island coastline.

ABOVE The Tui (*Prosthemadera novaeseelandiae*) is primarily a forest-dwelling species, though its range is spreading into settled habitats.

ABOVE North Island Weka (*Gallirallus australis greyi*) numbers have much dwindled in the past 150 years, but many survive on Kapiti Island.

BELOW A small honey-eater, very vulnerable to Ship Rat predation, the Stitchbird (*Notiomystis cincta*) survives on a few offshore islands like Kapiti.

BELOW The raucous call of the Kaka, closely related to the Kea but confined to a forest habitat, may be all that betrays its position.

Stephens Island (Takapourewa)

Stephens Island (Takapourewa) is an internationally important wildlife sanctuary. A tiny, windswept cliff-encircled island of only 150 hectares (370 acres), it lies three kilometres (1.9 miles) off the much larger D'Urville Island at the western extremity of the Marlborough Sounds. Its exceptional conservation interest relates to two characteristics: it has by far the largest Tuatara population of the 30 or so 'Tuatara islands', and the degree of endemism in its fauna is unusually high.

The Tuatara population is estimated to be above 30,000, and is as dense as 2,000 per hectare (810 per acre) in the coastal forest and shrubland. A small and extremely vulnerable population (about 300) of a separate species of Tuatara, the Brothers Tuatara, has been identified close by on the most inhospitable Cook Strait island – the North Brother Island, a four-hectare (ten-acre) rock on top of which the reptiles occupy a patch of scrub of only 1.7 hectares (2.9 acres).

Stephens Island is home to other interesting wildlife. There are three geckos, including the rare striped Stephens Island Gecko; four species of the skink genus *Leiolopisma*; the docile Stephens Island Giant Weta (which is actually smaller than the Stephens Island Tree Weta); and one of only two populations of the very rare Hamilton's Frog (further discussed in connection with Maud Island – see page 122). But there is a darker side to the history of the island, an all too typical example of the New Zealand tragedy of thoughtless introduction of mammalian competitors and predators to pristine island ecosystems. In 1894 a lighthouse was established on Stephens Island. Most of the forest was cleared, grazing stock was introduced, and – unrealized until after the event – the family's cats eliminated the vulnerable forest birds, including Kokako, Saddleback, a now extinct native thrush species and the Stephens Island Wren, once widespread also on the mainland. However, the Tuatara, lizards and many special invertebrates have survived, no doubt because the island is still rat-free.

ABOVE A basking Speckled Skink (*Leiolopisma infrapunctatum*). On Stephens Island they often shelter in Petrel burrows at night.

NEAR RIGHT The Marlborough Green Gecko (*Naultinus manukanus*) lives in the dwindling shrublands of the Marlborough Sounds. Yellow varieties also occur.

FAR RIGHT The Tuatara (*Sphenodon punctatus*), sole survivor of an ancient reptile group, the Rynchocephalia, otherwise extinct about 60 million years ago.

LEFT Rugged Stephens Island, the most exposed island in the outer Marlborough Sounds, is one of New Zealand's most important wildlife sanctuaries.

BOTTOM LEFT *Hebe urvilleana*, usually found associated with serpentine, is confined to a few islands in the Marlborough Sounds and the Nelson region.

BOTTOM CENTRE The bright red fruit of the Nikau Palm (*Rhopalostylis sapida*) is a favourite of native birds, especially New Zealand Pigeons.

BOTTOM RIGHT Once common on rocky coastlines, Cook's Scurvy Grass (*Lepidium oleraceum*) – used by Captain Cook as a remedy for scurvy – is now almost entirely confined to islands.

RIGHT The Wellington Weta (*Hemideina crassidens*), a tree weta (or 'bush weta')found throughout the Marlborough Sounds.

FAR RIGHT The Fairy Prion (*Pachyptila turtur*), New Zealand's smallest and commonest prion, often shares its burrows with Tuataras.

Maud Island

Maud Island is a rather nondescript island, 309 hectares (763 acres) in area, lying outside the entrance to Tennyson Inlet in the Marlborough Sounds. As with so many of the other topographical features of the sounds, the island is steep, with regenerating forest in the gullies but introduced pasture on the drier faces and ridge-crests. But it is the island's rodent- and cat-free status that has determined its selection as a sanctuary for two of New Zealand's most endangered flightless birds: the Kakapo and the Takahe.

Maud Island is also important for providing the largest habitat for the rarest of New Zealand's endemic frogs, the small, primitive Hamilton's Frog. This is confined to 15 hectares (37 acres) of remnant coastal forest (mainly Kohekohe and Mahoe). Like the other endemic species of the primitive *Leiopelma* genus, it does not require running water: its offspring do not go through a free-living tadpole stage, but are born as miniature frogs. Recent genetic typing of the Stephens Island and Maud Island frogs indicates that they are different species, giving the Stephens Island 'Hamilton's Frog' the dubious distinction of being the rarest frog in the world.

Also on Maud Island there is a small population of the world's rarest parrot, the Kakapo. Five of the remaining population (totalling fewer than 50 known birds) have been introduced to this haven from their last besieged natural strongholds in Fiordland and southern Stewart Island. The island is less than one kilometre (0.6 miles) from the mainland and hence vulnerable to reinvasion by stoats, which can swim this distance. Nevertheless, Maud Island is considered a key part of New Zealand's most urgent and intensive rare- and endangered-species recovery plan, the attempt to establish a captive-bred population of Kakapo in a semi-natural environment.

Maud Island is an important wildlife sanctuary, especially for Giant Weta, *Deinacrida rugosa* and small populations of Kakapo and Takahe.

ABOVE Looking east along Maud Island towards Pelorus Sound and the ridges around Mount Stokes. The island's core of broadleaf forest is the most important habitat for Hamilton's Frog, New Zealand's rarest native frog and the world's most primitive.

BELOW LEFT The continued sight of mature Southern Rata (*Metrosideros umbellata*) in flower, a red burst in the green canopy, is under threat as the tree is a favourite food of the introduced Brush-tailed Possum.

BELOW Hamilton's Frog (*Leiopelma hamiltoni*), found only on Maud and Stephens islands; the small population on the latter may be a separate species. They have limited vocal ability, live in a moist, misty environment and protect their eggs and young.

Farewell Spit

Farewell Spit is the northernmost part of the South Island. One kilometre (0.6 miles) in width, it is a classic sandspit of international significance, a wide arc of sand that trails off eastwards for nearly 30 kilometres (19 miles) from Cape Farewell into Golden Bay. It is by far the largest spit landform in New Zealand, and has the most extensive sand dunes in Nelson. In the spit's lee lie 10,000 hectares (38.6 square miles) of intertidal sandflats, formed by westerly winds constantly sweeping the sand that accumulates on the spit eastwards into the shallows of Golden Bay. These intertidal sandflats are what makes Farewell Spit such an outstanding wildlife habitat. The whole ecosystem – spit and intertidal flats – is protected as a nature reserve, the strictest category of conservation protection. In addition, the area has been designated a Wetland of International Importance under the UN Ramsar Convention.

Over 80 species of bird frequent Farewell Spit, making it internationally famous in ornithological circles. Most are shore birds, mainly migratory waders that breed in Arctic and subarctic regions. Three species – Godwit, Knot and Turnstone – make up over 90 per cent of these, with about 45,000 flocking to the area between spring and autumn. They include small numbers of migrants not well known to most New Zealanders, such as the Hudsonian Godwit, Siberian Tatler and Mongolian Dotterel.

The spit is also very important as a feeding ground for many common New Zealand waders. These species, like the Pied Oystercatcher, Banded Dotterel and Pied Stilt, breed in the braided riverbeds of the harsh high country of inland Canterbury and Otago, but prefer to spend the autumn and winter in the mild climes of Golden Bay.

Looking south-west from the Farewell Spit dunes across the shallow intertidal flats of Golden Bay to the hills of the Wakamarama Range.

ABOVE Australasian Gannets are on the increase in New Zealand, the colony at Farewell Spit being the most recent mainland site adopted by the birds. By the end of May the chicks have usually fledged and left the colony, which is then deserted until the next breeding season.
RIGHT Most of Farewell Spit is only about a kilometre (just over half a mile) wide. At Puponga Farm Park, near its base, is an intricate landscape of dunes and dune lakes. From here one can observe prions, White-fronted Terns and Blue Penguins near Wharariki Beach and Fossil Point. Pilot Whales are regularly beached in the shallows of Golden Bay.
BELOW New Zealand's commonest migrant Arctic wader, the Eastern Bar-tailed Godwit (*Limosa lapponica baueri*). All told, about 100,000 godwits migrate annually to New Zealand, most frequenting the estuaries of Northland and the Firth of Thames.

Nelson Lakes National Park

This park gets its name from the twin lakes Rotoiti and Rotoroa, which occupy old glacial troughs in the mountainous headwaters of the Buller River; they are by far the most significant freshwater bodies in a region lacking in lakes. Both are accessible from State Highway 6 between Nelson and the West Coast, and each is cold and crystal clear, nestling in a picturesque setting of (beech) forest-covered mountains.

The park straddles the Spenser Mountains at the northern end of the Southern Alps. These mountains are almost uniformly of greywacke, a hard but brittle type of sandstone that makes up much of the main mountain chain of the South Island; further uniformity is provided by the similarity in height of all the ridge-crests – 2,200-2,300 metres (7,200-7,500 feet) – and the presence of all four species of Southern Beech (*Nothofagus*). Beech forest entirely dominates the landscape. Red Beech and Silver Beech share the lower, warmer and more fertile sites; Hard Beech can be found on the poorer soils at lower altitudes; and the hardy Mountain Beech (with Silver Beech in the wetter regions) extends upslope to the tree-line. The tree-line itself is one of the truly remarkable features of the park – again for its uniformity. It is almost as if a draughtsman had drawn a straight line across the valley sides at about 1,400 metres (4,590 feet).

But this lack of diversity does not mean drabness. There is a sparkle to the filtered light of beech forests, making them so much more welcoming than the dense podocarp rainforest of the lowlands. Above the bush-line there are colourful fell-fields, full of alpine flowers, Snow Tussocks and speargrass. The forests are frequented by insect-eating birds – especially smaller birds like Riflemen (one of the smallest of wrens), South Island Robins, Tomtits, fantails and Grey Warblers.

BELOW The forests of Nelson Lakes National Park are almost entirely Red, Silver, Hard and Mountain Beech.

BOTTOM LEFT The hoods of Greenhooded Orchids (*Pterostylis* spp) have a hinged lip to flip insects into the hood.

BOTTOM CENTRE *Gastrodia cunninghamii*, one of New Zealand's three 'potato orchids'. Lacking leaves and chlorophyll, they draw nutrients from tree roots via fungal threads.

BOTTOM RIGHT The South Island Robin (*Petroica australis*), a charming opportunist that will feed off forest-floor invertebrates stirred up by visitors.

BELOW The tiny Rifleman (*Acanthisitta chloris*) is often found in small flocks in the park, methodically picking insects from the trunks of the beech trees.

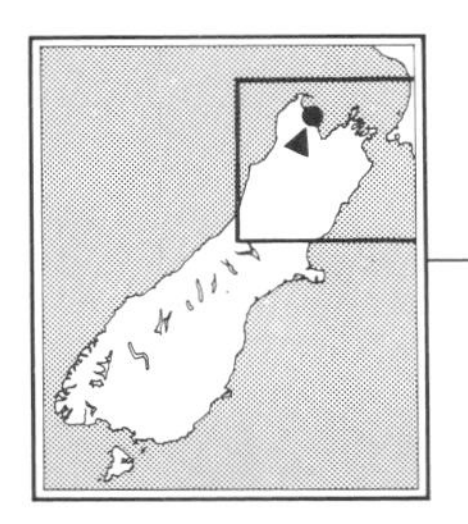

Abel Tasman National Park

This is the smallest of New Zealand's National Parks, at 22,500 hectares (87 square miles); it is also one of the few that have been seriously influenced by fire and introduced plants. Its interior is very rugged, consisting almost entirely of the Pikikiruna Range, between Golden and Tasman Bays. To most visitors Abel Tasman National Park is synonymous with the beautiful golden sand beaches and granite headlands along the three-day Abel Tasman Coastal Track. The coastline is subjected to one of the largest tidal ranges in the country – up to four metres (13 feet). This energy sweeps the sand into a 'longshore drift', forming delicate and intricate networks of bars and spits at places like Torrent Bay, Bark Bay, Awaroa Inlet and Wainui Inlet. The coastline is a gallery displaying nature's artistic ability; at every turn there are wonderful original sculptures in the granite bedrock. In places the coastal waters are dotted with small islands, each with their characteristic apron of tan-coloured granite boulders, a platform just above sea level which has been formed by the waves constantly undercutting the rock pedestal. These wave-cut platforms are a favourite roost for New Zealand Fur Seal colonies, which thrive on Tonga and Adele islands. The islands also have good populations of Little Blue Penguin, survivors of those largely wiped out on the park mainland through predation by stoats.

The karst landscapes of the Pikikiruna Range and the Canaan tableland provide plenty of challenge for the caver, especially the abseil down 176 metres (577 feet) into the awesome mouth of Harwoods Hole, the deepest vertical cave-shaft in the country. The uplands are heavily forested in Red, Silver and Mountain Beech. An interesting botanical feature of this curious little park is that it hosts all four species of beech – as well as the 'Black Beech' subspecies of Mountain Beech.

Torrent Bay, one of many idyllic bays in Abel Tasman National Park. The several-metre tidal range around here is among New Zealand's highest.

ABOVE Slaughter for their skins led to the disappearance, last century, of New Zealand Fur Seals (*Arctocephalus forsteri*) from many parts of the coast. Now protected, they are gradually returning.

LEFT The Anchorage, near Torrent Bay, one of the largest beaches in Abel Tasman National Park. The park takes its name from the Dutch navigator Abel Tasman (1603–*c* 1659), who in 1642 was the first European to sight 'Neuw Zeeland'.

LEFT Frequent tree ferns show the forests of Abel Tasman National Park are healing themselves after the fires that often swept this coastal landscape before it was designated a National Park in 1942. Beech (Hard, Black, Red and Silver) forest survives in the moister coastal parts of the park, where it is more suited for survival on the steep soils on the deeply weathered granite.

BELOW Waikoropupu Springs, close to Takaka in Golden Bay. Cold, extremely pure water percolates down from karst within Abel Tasman National Park to bubble spectacularly to the surface here. The clarity is the highest measured for a freshwater system anywhere.

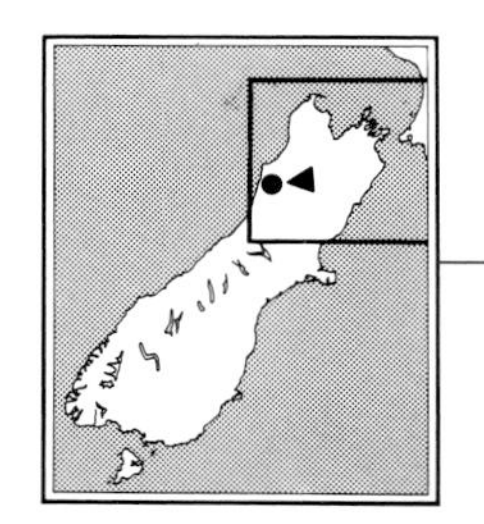

Paparoa National Park

Paparoa National Park was formed in 1987 after a bitter campaign to protect its forests and minerals from exploitation; it was the first of New Zealand's modern 'scientific' parks, its boundaries being chosen carefully to represent the full range of ecosystems. In just 30,000 hectares (120 square miles) or so of very wild and largely untracked country the landscape contrasts are amazing: the gendarmes and spires of the narrow crest of the Paparoa Range; the canyons of Bullock Creek and the Fox and Pororari rivers; the delicate limestone formations of the Ananui (Metro) Caves; the wild surf, cliffs and stacks of the coastline between Seal Island and Razorback Point; and, above all, the fascination of the dramatic performances of nature at the 'Pancake Rocks' on Dolomite Point. Here the pounding of the waves on the chasms, pipes and platy formations of the coastal limestone produces blowholes – huge waterspouts which rival their thermal counterparts in the geyserlands of Rotorua.

The park's karst landforms are clothed in lowland rainforest *par excellence*. The luxuriance of the rainforest makes it seem subtropical, with a profusion of Nikau Palms and tall, slender black Mamaku Tree Ferns. Huge Northern Rata trees thrive in this mild and humid climate, towering along with Rimu and Miro above the canopy of densely understoreyed hardwood trees. Inland, where cold mountain air descends to form pockets in the basins, beech forest is more prevalent.

Forest birds such as Tui, Bellbird, pigeon, Kaka and parakeet migrate seasonally between the coastal and upland forests. The park is an important habitat for Great Spotted Kiwi, heard but seldom seen because of its nocturnal habits and the difficult terrain. Another twilight bird in the park is the endemic Westland Black Petrel, which breeds only in the coastal forests between Barrytown and Punakaiki.

The 'Pancake Rocks' and the Tasman Sea's wild surf at Dolomite Point near Punakaiki on Paparoa National Park's coast.

LEFT Paparoa National Park's lowland coastal rainforests are particularly luxuriant; tall tree ferns (Mamaku) and dense groves of Nikau Palm give them a subtropical character. At the base of the limestone cliffs, huge Northern Rata, Rimu and Miro tower above a 'jungle' of broadleaf trees, ferns, epiphytes and lianas (lianes) like Supplejack.

BELOW The Westland Black Petrel (*Procellaria westlandica*), an endemic species that breeds only on the cliffs between the Punakaiki River (in Paparoa National Park) and Lawson Creek. The colony, not discovered until 1945, is unusual in being one of the few petrel breeding colonies still on the mainland, a tribute to the birds' ability to hang on despite rats, cats, dogs, and stoats.

BELOW Once hunted for food and feathers, the Western Weka (*Gallirallus australis australis*), found mainly in the Nelson region and North Westland, is now protected. Wekas, notorious for scavenging, are flightless, but their strong beak and legs protect against predation.

BELOW The Great Spotted Kiwi or Roa (*Apteryx haastii*), one of the larger Kiwis, weighing up to 3.7 kilograms (just over eight pounds), is widely distributed in the forests of Kahurangi National Park, Paparoa National Park and other parts of North Westland. Rarely seen despite their size, they may be heard at night making either a loud snuffling or a series of harsh whistling calls.

The best-known New Zealand Fur Seal rookery on the South Island's western coast lies close to Cape Foulwind, near Westport. The rookery is a major tourist attraction, as it lies only ten minutes' walk from the road end in Tauranga Bay.

Kaikoura

Kaikoura has always attracted fishers, Maori and Pakeha alike; *kai koura*, meaning 'to eat crayfish', indicates how the area has always been important to Maori people. Today Kaikoura and its surroundings waters have become New Zealand's supreme location for the observation of coastal wildlife.

Several environmental factors combine to make the Kaikoura Peninsula such an outstanding wildlife habitat. The limestone and siltstone of the peninsula have been eroded by the sea into an extensive offshore reef and a wide variety of exposed or rocky shores. The Kaikoura Walkway, running around the tip of the peninsula, traverses these habitats and offers a vantage point from which to view the breeding populations of New Zealand Fur Seal, Red-billed Gulls and White-fronted Terns. The diversity of marine life is impressive and of enormous recreational and educational value; there are over 120 species of fish and macro-invertebrates, 45 species of seaweed and over 60 species of mollusc.

But whale-watching is the premier wildlife attraction. Why do adolescent male Sperm Whales congregate in this place, routinely coming much closer to shore than in most other parts of the world? The answer is a combination of underwater topography and the mixing of cold and warm currents off the peninsula's shore. The deep Hikurangi Trench – extending all the way south from the subtropical waters of the Kermadec Trench – terminates in the Kaikoura Canyon, some 1,500 metres (4,920 feet) deep and a mere five kilometres (three miles) or so offshore. Here, the whales dive to more than 1,000 metres (3,280 feet) in search of their favourite food – Giant Squid. After feeding in the canyon, the Sperm Whales surface among Orca (Killer Whales), dolphins and seals in Kaikoura's coastal waters within easy reach of carefully regulated observation boats.

BELOW Winter view across South Bay on the Kaikoura Peninsula towards the Seaward Kaikoura Range, with Southern Black-backed Gulls (*Larus dominicanus*) foraging among the rocks.

RIGHT The nocturnal Black-eyed Gecko (*Hoplodactylus kahutarae*), one of New Zealand's rarest geckos, is confined to rocky alpine outcrops in the Seaward and Inland Kaikoura ranges. No other New Zealand lizard is known to live at such a high altitude.

CENTRE RIGHT Dusky Dolphin (*Lagenorhynchus obscurus*), the most abundant and spectacular performers of the dolphins frequenting the Kaikoura coast; Hector's Dolphins are another visitor. A marine-reserve proposal focuses on the extensive reefs and rocky platforms off the end of the Kaikoura Peninsula.

BELOW Kaikoura is one of the best places in the world to view whales, especially the Sperm Whale (*Physeter catodon*). This one is starting a dive, probably to 1,000–2,000 metres (3,300–6,600 feet) in search of its favourite prey, Giant Squid. The largest toothed whale, the Sperm Whale can reach 20 metres (66 feet) in length.

THE SOUTHERN ALPS AND WESTLAND

The steepness and beauty of the Southern Alps astonish first-time visitors to Canterbury or the West Coast. Their snow-covered heights dominate the skyline of the central part of the South Island, extending in an unbroken north-east/south-west line for 450 kilometres (280 miles), from Lewis Pass to Key Summit Divide on the Milford Road. In the central portion, from Whitcombe Pass in the head of the Rakaia River to Haast Pass, the peaks are generally above 2,500 metres (8,200 feet); in Mount Cook National Park the main divide peaks tower 3,000–3,700 metres (9,800–12,100 feet) above sea level. It was these peaks 'of prodigious height' that led Captain James Cook, sailing along the West Coast during his first visit to New Zealand in 1770, to name the range the Southern Alps.

These mountains are of deep significance to the Ngai Tahu, whose tradition reveres them as ancestors. In their mythology the South Island is the stranded canoe of Aoraki, the eldest son of Raki the Sky Father. Aoraki and his brothers scrambled onto the high side of the canoe, and became the magnificent mountains of the Southern Alps. Aoraki itself, at 3,754 metres (12,316 feet) the highest of them all, was subsequently named Mount Cook in honour of the great English navigator, and today both names tend to be used interchangeably.

To the east and west of this great mountain chain lie wild lands carrying the imprint of the glaciers that extended far out from the Southern Alps during the last ice age. To the east stretch the vast open valleys and intermontane basins of inland Canterbury and the upper Waitaki. This 'high country' is a land of *Festuca* tussock grasslands and Matagouri, nor'-west winds and braided rivers, Black Stilts and soaring Harrier Hawks. Its tan-coloured grasslands are broken only by the azure blueness of the chain of great glacial lakes extending from Lake Coleridge – 80 kilometres (50 miles) west of Christchurch – to Lake Wakatipu in the Otago lake country. In contrast, the west is a land of tall dense rainforest, with wild and remote mountain gorges, intense rainfall – measuring as high as 13,000 millimetres (512 inches) annually – and a string of swamps and slow-flowing creeks along the narrow coastal plain. So impressive is this land of mountains, lakes and rainforest that almost all of it has been protected as National Parks. Arthur's Pass was the first (1929), followed by Mount Cook (1953), Westland (1960) and Mount Aspiring (1964). In addition, 40,000 hectares (154 square miles) of the wild and remote mountains around Mount Hooker and the Landsborough Valley in South Westland are managed as a wilderness area – that is, in a wild state with no tracks, huts, vehicles or air access.

The most significant forest conservation achievement of the last decade occurred in this region. In 1989 all the forests south of the Cook River – over 300,000 hectares (1,160 square miles) – were protected as conservation land in a controversial political decision to stop the logging of their timbers. These lowland podocarp forests of South Westland are internationally valued as the largest

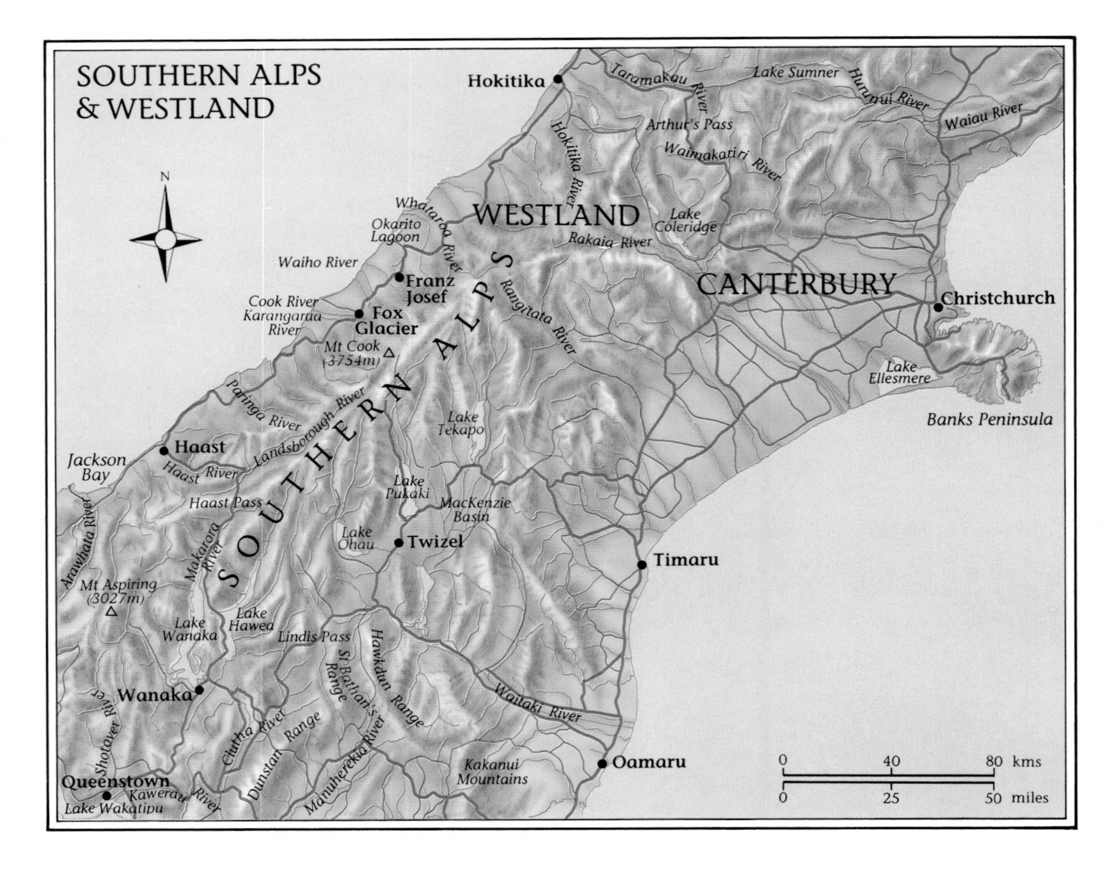

ABOVE The endemic New Zealand Falcon (*Falco novaeseelandiae*), much at home in Southern Alps beech forests, is smaller and darker than the other native raptor, the Australasian Harrier (*Circus aproximans*). The eggs in its nest, usually just a scrape under a rock ledge, are orange-brown.

RIGHT The tussocklands of the South Island high country, in the Southern Alps' rain shadow, extend as a belt of native grasslands from Molesworth to the Takitimu Mountains.

remnant linking modern New Zealand to the great Mesozoic swamp forests of the ancient supercontinent Gondwana (Gondwanaland). Their protection allowed UNESCO, in 1990, to designate the South Island's vast south-west sector – 2.6 million hectares (10,000 square miles) – as the Te Wahipounamu World Heritage Area.

The scenery of the region is world-class, a product of the interplay of huge climatic and geological forces. The Alps are a barrier wall standing against the moisture-laden westerly winds driving in from the Tasman Sea. The winds drop most of their moisture as rain on the lowlands and foothills of the West Coast or as snow when they are pushed up over the peaks. By the time they descend to the rain shadow of the eastern basins they are often dry and warm -- classic foehn winds that for millennia have scoured the braided riverbeds, lifting clouds of glacial silt and depositing it as loess on the surrounding landscape.

The Southern Alps are among the most dynamic, and dangerous, regions in New Zealand. The Alpine Fault, a huge rent in the Earth's crust, lies at the foot of the mountain belt, about 15–25 kilometres (9–15 miles) west of the range's crest. Here not only are the Pacific and Indian-Australian tectonic plates (huge continent-sized pieces of the Earth's crust) grinding past each other along the fault, but the Pacific Plate is being pushed up over the other to form the Southern Alps. Uplift has accelerated during the past five million years, and today the rate of uplift of Mount Cook is five millimetres (0.3 inches) per year, while close to the Alpine Fault the lower slopes are rising at an impressive 10–20 millimetres (0.4–0.8 inches) annually. Geologists estimate that a total depth of 25 kilometres (15 miles) of greywacke and argillite rock may have been uplifted during the formation of the Southern Alps. However, the mountains are considered to have kept the same height throughout the Pleistocene ice age, indicating that this extraordinarily high rate of mountain-building was (and is) matched by a similar rate of natural erosion. A very spectacular illustration of this uplift/erosion equilibrium was the well documented Mount Cook rock avalanche of December 14, 1991.

During the night the summit rock and icecap collapsed, plummeting an estimated 14 million cubic metres (493 million cubic feet) of debris down to the surface of the Tasman Glacier, 2,700 metres (8,850 feet) below, at speeds approaching 600 kilometres (370 miles) per hour, and instantly reducing the height of Mount Cook by ten metres (33 feet). The momentum of the avalanche front, two kilometres (1.25 miles) wide, carried it across the 1.5-kilometre (0.9-mile) Tasman Glacier and 70 metres (230 feet) up the slopes of the Malte Brun Range on the far side – 7.5 kilometres (4.7 miles) from its point of origin.

Glaciers, hundreds of them, are spread throughout the Southern Alps. The region's only glaciers outside the Alps are the impressive but small shelf glaciers around Milford Sound in Fiordland and the fast-disappearing glacial remnants on Mount Ruapehu. The modern glaciers of the Alps, though likewise only remnants – fragments of the huge ice-sheets and valley glaciers of the ice age -- are some of the most accessible and impressive of the world's temperate glaciers. The Tasman Glacier is, at 28 kilometres (17 miles), the longest in New Zealand, its upper reaches providing unrivalled experiences for those who 'ski the roof of New Zealand'. On the western side of the Alps, the Franz Josef and Fox glaciers are internationally famous tourist attractions: they are steep and narrow, with impressive rapid-flowing icefalls that terminate at an altitude below 300 metres (980 feet) in the rainforest zone. Other notable glaciers are the Garden of Eden and Garden of Allah, north of Mount Cook National Park, and the névés around Mount Aspiring and the Olivine Ice Plateau in the heart of the alpine wilderness that is Mount Aspiring National Park.

The legacy of glaciation is so extensive that virtually every landform in the region – and hence the soils, vegetation and, indirectly, wildlife -- is a result of some glacial influence. The only real exceptions are the landforms subsequently built by rivers and the ever-present wind, or by the rising of the sea level on the western coastline after the last glaciation. Everywhere at the foot of the mountains, east and west, lies the handiwork of glaciers and meltwaters: smooth outwash plains and terraces, strands of moraine, lakes and tarns filling hollows once occupied by stagnant ice (kettleholes), or isolated rock monoliths with their faces scoured and plucked by the passing ice (roches moutonnées). The mountains, too, are replete with the imprint of ice – arêtes and horns, cirques and hanging valleys: in all, a geomorphologist's dream. Nowhere else can the visitor get such a graphic insight into the healing power of nature as on the surfaces bared by the retreat of the Franz Josef and Fox glaciers. The Franz Josef soil 'chronosequence' is the best studied, revealing natural colonization in a mere 20 years by nitrogen-fixing herbs and shrubs, a Rata-Kamahi forest grown to a height of 20 metres (65 feet) in 150 years, and tall podocarp trees – Rimu, Miro and Hall's Totara – developed after about 1,000 years.

With such a profusion of glacial landforms and soil ages, there is a wide diversity of plant-life, particularly in Westland National Park, where there is a complete altitudinal sequence from the crest of the Southern Alps – 3,497 metres (11,466 feet) at Mount Tasman – to sea level. While beech forests and short tussock grasslands cover most of the mountain landscapes in the eastern rain shadow of the Alps, the west is the realm of the great lowland podocarp rainforests of New Zealand. Three distinctive types of podocarp forest stand out: the dense Rimu forests of the terraces and moraines; the Kahikatea forests of the fertile alluvial floodplains and swamp margins of South Westland; and the colourful 'shrub podocarps' (such as Silver Pine and Bog Pine) of the extremely infertile wet soils on the oldest surfaces, like the Waikukupa high terraces in Westland National Park.

The Rimu forests were the mainstay of the West Coast timber industry for over a century, but public outcry from the mid-1970s onwards over the failure of the industry to manage them in a sustainable fashion ultimately led to the protection of all such forests south of Okarito Lagoon. The slender and graceful Kahikatea is New Zealand's tallest tree and also the 'oldest', in the sense that it was the first podocarp to appear in the fossil record. A botanical curiosity of the shrub-podocarp community is the world's smallest conifer, the Pygmy Pine (*Lepidothamnus laxifolius*).

Because the Southern Alps and the West Coast are still overwhelmingly natural in character, they present one of the most important wildlife habitats in the country. The braided riverbeds of the upper Waitaki River catchment are the home of two distinctive endemic birds: the Wrybill, with its curious beak curved to the right, and the very rare and endangered Black Stilt. Arthur's

Lake Tekapo, one of the great glacial lakes of the South Island high-country tussock grasslands. With its sister lakes, Pukaki and Ohau, it marks the troughs formed by the ice-age glaciers that advanced out of the Southern Alps around Mount Cook. The exotic trees in the foreground shelter Lake Tekapo village from the north-westerly gales that sweep down the Godley Valley at the head of the lake.

Pass National Park has a Great Spotted Kiwi population, although no Brown Kiwi. The Alps have the most significant populations of New Zealand's engaging mountain birds: the Kea (the only alpine parrot in the world), the tiny Rock Wren (an endemic member of an ancient genus, *Xenicus*), and the New Zealand Falcon. The amusing antics of the Kea are legion and a delight to all visitors, except those whose cars or tents are damaged by these incorrigible mischief-makers. The Rock Wren is no less pleasing, with its excited bobbing dance, but it is elusive and generally seen only by mountaineers.

The hanging valleys and gorges of South Westland are a stronghold for Blue Duck (Whio), and the area contains the most important populations of increasingly rare forest birds like South Island Kaka, Rifleman, Yellowhead and Yellow-crowned Parakeet. Okarito Lagoon, adjacent to Westland National Park and the largest estuarine lagoon on the West Coast, is home to Pied Oystercatchers and many other wading birds. Close by lies New Zealand's single nesting colony of White Heron (Kotuku); the 60 or so pairs that congregate in a few Kahikatea trees in Waitangiroto Nature Reserve provide a wonderful sight, evoking the almost mystical value that the bird has for Maori people, symbolizing all that is beautiful, rare and worthy of protection.

Arthur's Pass National Park

Arthur's Pass National Park straddles the Southern Alps between the Waimakariri River in the east and the Taramakau in the west. It is rather uniform geologically (greywacke/argillite), mountainous and notable for its many huge rockfalls, the largest of which extends off Falling Mountain for three kilometres (1.9 miles) down the Otehake Valley. Lying 100 kilometres (60 miles) from Christchurch, the park is very accessible via either the Tranz Alpine Express on the Midland Railway or State Highway 73 across Arthur's Pass to the West Coast. The completion of the Otira Tunnel to the West Coast in 1923 allowed regular train traffic and hence recreational visits to Arthur's Pass village by the people of Christchurch. Concern about the area's natural environment led to its protection as a National Park in 1929, the first of a chain of such parks along the Southern Alps.

The park has a simple charm: peaks about 2,000-2,200 metres (6,600-7,200 feet) in height, extensive scree slopes of fractured greywacke, wide braided rivers in the east, steep gorges in the west, and ten small glaciers perched high on the main peaks above the Bealey and Waimakariri valleys. The east-west contrast in forest type is marked. Mountain Beech forest is so widespread on the eastern slopes that it is almost a monoculture, an unbroken green cloak of constant wove extending from the river's edge to the bush-line. The western forests are far more complex: podocarps and tall Red and Silver Beech occupy the valley floors; Kamahi, Southern Rata and Hall's Totara are on the middle slopes; and, higher still, there are Kaikawaka (Mountain Cedar) and the colourful but near-impenetrable subalpine shrub belt which replaces the clean beech bush-line of the east. Alpine flowers -- buttercups and gentians, eyebrights and celmisias – grow in profusion from October to February. These, the visual highlight of the park, can be easily viewed from the Dobson Nature Walk on Arthur's Pass.

Mount Rolleston, the most accessible of the higher peaks in Arthur's Pass National Park, is only three kilometres (two miles) from Arthur's Pass.

ABOVE In the snow tussock herb-fields of Temple Basin, above the 920-metre (3,018-foot) Arthur's Pass, alpine flowers grow in profusion in late spring. Here the world's largest buttercup, the striking Mount Cook Buttercup (*Ranunculus lyallii*), grows in clusters along the margins of a small mountain stream.

RIGHT Juveniles of the Kea (*Nestor notabilis*) can be identified by the yellow feathers around the beak and nostrils.

FAR RIGHT Clumps of *Celmisia* in Arthur's Pass National Park. Some 50 of the 60 species in the large Australasian genus *Celmisia*, part of the Daisy family, occur in alpine country.

Mount Cook National Park

Mount Cook National Park is New Zealand's great alpine park. The Southern Alps rear abruptly above the tussock basins of the upper Waitaki River, the so-called Mackenzie Country, named for James McKenzie, a notorious sheep rustler who became part of the folklore of the High Country.

The park is a harsh land of rock and ice, with 40 per cent of its area still covered by glacial ice. Five large glaciers fill the upper reaches of the main valleys: the Mueller, the Hooker, the Tasman and its tributary the Murchison, and the Godley in the remote north-east at the head of Lake Tekapo. There are 19 peaks over 3,000 metres (9,800 feet) in height ranged along the main divide from Elie de Beaumont to Sefton and in the adjacent Mount Cook and Malte Brun ranges. They provide a wonderful variety of high climbing and skiing challenges, unparalleled elsewhere in the Southern Alps.

The mountainous setting of the approach along State Highway 80 is magnificent: the milky blueness of Lake Pukaki, the short tussocks and Matagouri, the silvery braids of the Tasman River flowing into the lake, and the fantastic cloud arches that so often herald the approach of a nor'west storm. The highway ends at Mount Cook village and the park visitor centre, just below the terminals of the Hooker and Mueller glaciers. There are breathtaking views of Mount Cook above the Hooker Valley and of Mount Sefton seemingly only a stone's throw away across the Mueller Glacier. The village is dwarfed by the scale of the mountains and always at risk from the storms that can scourge the park, unabated, for two weeks at a time.

There is hardly any forest, save small pockets of Totara and Silver Beech in the lower valleys. Instead, the park is alive with the most wonderful variety of alpine plants: meadows of Mount Cook Buttercups, spiky Wild Spaniards, woolly cushions of 'vegetable sheep' on greywacke rubble, and delightful patches of South Island Edelweiss high above the snow-line.

Short tussock grasslands around lakes Tekapo and Pukaki lie in areas the Mount Cook National Park glaciers covered during the last ice age.

TOP The New Zealand Pipit (*Anthus novaeseelandiae*), widespread in the high country.

ABOVE The Black Stilt (*Himantopus novaezealandiae*), one of New Zealand's rarest birds, is restricted to the braided riverbed of the upper Waitaki.

RIGHT The Golden Spaniard (*Aciphylla aurea*), found in drier tussock grasslands, is tolerant of fire and grazing.

BELOW Weeds like this Dock (*Rumex* spp) are invading Mount Cook National Park.

BELOW RIGHT The Snowbell (*Gaultheria crassa*) can reach one metre (40 inches) in height.

ABOVE The Hooker River, one of the short, wild rivers issuing from Mount Cook National Park's glaciers, is always turbid with suspended glacier-ground 'rock flour'. On the skyline is Mount Sefton (3,157 metres [10,358 feet]).

BELOW Just outside the park, near the Glentanner high-country sheep station, the Tasman River flows into Lake Pukaki.

RIGHT Mount Tasman is, at 3,498 metres (11,476 feet), the highest mountain on the main divide of the Southern Alps. Its beautiful ice peak is obscured behind Mount Cook (lying just south, in the Mount Cook Range) in the famous view up the Hooker Valley from the Hermitage Hotel (see picture on page 15).

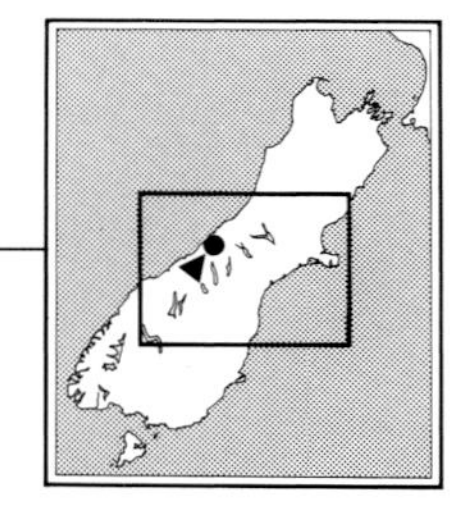

Westland National Park

Westland National Park is no less a landscape of superlatives than Mount Cook. The two share a common boundary, 40 kilometres (25 miles) along the highest crests of the Southern Alps, but to most visitors they are worlds apart. Westland National Park has high peaks and over 60 glaciers, but it includes also coastal lands along the Tasman Sea; in 1982 the lowland terrace Rimu of Okarito and Waikukupa forests were added after a campaign to protect them from logging. The park now fully represents the landforms and vegetation types on the wet western side of the Alps: it is New Zealand's 'mountains to the sea' landscape, a park of dramatic contrasts.

The Alpine Fault divides Westland National Park in two. East of the fault rise sheer forested slopes rent by deep gorges impassable to all except the hardiest mountaineer. High above, out of sight, nestle the myriad névés that feed the enormous descending trunks of the Fox and Franz Josef glaciers. To the west lies a largely untracked wilderness of dense rainforest, the undisputed realm of the podocarp family: the soaring ranks of Rimu crowding the narrow road to Gillespies Beach, the stately groves of Matai and Totara on the Cook River Flats, the colourful Yellow-silver Pine and Bog Pine of the infertile and wet plateau around Lake Gault, and the conical symmetry of the Kahikatea fringing Lake Mapourika and Lake Matheson. To watch from the coast at Okarito Lagoon or the mouth of the Waiho River as the sun sets on these forests and the snows of the Southern Alps is to be one with a great work of nature.

Because of the relatively mild climate and the diversity of lakes and wetlands, the park's lowlands are an outstanding habitat for aquatic wildlife. The most aquatic of New Zealand's freshwater birds, the Southern Crested Grebe, breeds here. New Zealand Scaup, Pukeko, Royal Spoonbill, White Heron, Fernbird and New Zealand Kingfisher can be watched in the estuaries and forest-fringed lagoons.

On a clear day up-valley by the Fox River, the prominent snowy peak of Mount Tasman stands out. Like most rivers in the park, the Fox tumbles steeply out of the Southern Alps, crossing the Alpine Fault at this point.

ABOVE Lake Mapourika, a scenic gem of Westland National Park. State Highway 6 runs by its eastern shores, affording fine views of the flax and Kahikatea forest fringing the water.

BELOW LEFT The Franz Josef Glacier's terminal face. The thin ice ribbon at the face (and in the icefall above) reflects climatic changes governing the glacier's growth or decrease.

BELOW The lower reaches of the Fox Glacier icefall are a jumbled mass of séracs (ice pinnacles) and crevasses. Tourist access to the terminal is often blocked by rockfall or river erosion.

LEFT Kotuku, the White Heron (*Egretta alba modesta*), most strikingly beautiful of wetland birds, is found in Australasia and Asia. The only place it breeds in New Zealand is in a few trees of the Waitangiroto River's Kahikatea forest, beside Westland National Park, where there are now about 120 birds.

BELOW LEFT Because introduced possum have not yet penetrated far into South Westland, Southern Rata (*Metrosideros umbellata*), with its crimson flowers, is still a common component of forests in Westland National Park and South Westland. It is smaller than Northern Rata (*Metrosideros robusta*) and unlike it does not begin life as an epiphyte.

BELOW The White Heron colony in Waitangiroto Lagoon near Okarito. The large birds are dwarfed by the huge old Kahikatea (*Dacrycarpus dacrydiodes*) trees, festooned with epiphytes and ferns. Flax fringes the lagoon's deep waters, stained deep brown by humic substances leached from the organic soils of this 'floating forest'.

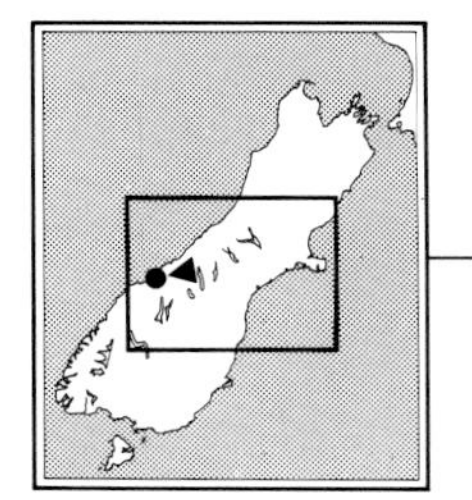

South Westland and the Haast Coastline

South of the Cook River and Westland National Park lie the West Coast's most remote and wild coastlines. New Zealand's premier scenic forest-heritage highway, State Highway 6, threads its way along the Alpine Fault, touching the coast briefly at historic Bruce Bay and crossing wild river after wild river – Karangarua, Makawhio, Mahitahi, Paringa, Moeraki, Waita and, finally, the largest of them all: the Haast. The Paringa supplies a significant biogeographical boundary, for here the familiar beech forests appear again after being absent from all of central Westland. Why there are no beech forests for 200 kilometres (125 miles) between the Taramakau and Paringa rivers is one of many forest riddles on the West Coast.

At Ship Creek, State Highway 6 descends to the Haast coastal plain, one of the country's outstanding natural areas. Ranks of windshorn Rimu stand fast on the foredunes, braced to withstand the constant assault of the westerly winds. The Ship Creek nature walks are world-class, two of a series of interpretative walks for visitors developed on the coastal plain by the Department of Conservation after the Government's 1989 decision not to allow any logging of public forests in South Westland. The walks meander among the dune lakes, Kahikatea swamp forest and Rimu-covered dry sand-dune ridges; from the air the plain is a wonderful maze of parallel forested dunes interspersed with elongated lakes – all from sediment deposited by the Haast River over the last 8,000 years.

These forests and wetlands are the core of the Te Wahipounamu World Heritage Area, New Zealand's most tangible link to the ancient supercontinent Gondwana; their story is graphically told at the Haast World Heritage Visitor Centre. South of Haast, the Hapuka Estuary boardwalk takes the visitor easily through the fringing Kowhai forest, with panels on the boardwalk telling the story of another famous inhabitant of the West Coast: Whitebait (or Inanga), a renowned delicacy.

The rocky shoreline south of Lake Moeraki, near Knights Point, is an ideal breeding-ground for Fiordland Crested Penguin and New Zealand Fur Seal. This is South Westland's most dramatic stretch of coast.

ABOVE Mataketake dune lake, near Ship Creek, one of a series of forest-fringed dune lakes parallel with the coastline. These, the outstanding forested dunes (with associated wetlands) in New Zealand, have formed over the past 6,000 years as the coastline has been pushed westwards by sediments washed down from the mountains.

LEFT Lake Moeraki, the last of the forest lakes beside the scenic highway from Hokitika to Haast. The lake lies just south of the 'beech gap', so its forests contain both beeches and podocarps. The Monro Walk leads from the lake's outlet through this forest to the coast at Monro Beach. Native bats and Fiordland Crested Penguin (*Eudyptes pachyrhynchus*) frequent this isolated coastline.

BELOW Often throughout coastal South Westland, windshorn Rimu forest comes right to the high-water mark; this magnificent stand runs several kilometres alongside Bruce Bay. The traditional Maori name for Bruce Bay is Mahitahi; recent archaeological investigation here and elsewhere in South Westland shows this coast once had a significant Maori population.

RIGHT The Arawhata, one of the wildest and largest of the many rivers draining the mountains of South Westland, rises in the wilderness heart of Mount Aspiring National Park.

Mount Aspiring National Park

At 3,027 metres (9,925 feet), Mount Aspiring is the only peak of over 3,000 metres outside Mount Cook National Park. It is a gleaming white horn, reminiscent of the European Alps' famous Matterhorn, and in its symmetrical beauty it vies with Mount Taranaki as New Zealand's most attractive mountain. Although it is the most prominent peak in the vast – 355,000 hectares (1,370 square miles) – park that bears its name, straddling the southern end of the Alps between north-west Otago and South Westland, there are hundreds of others.

This is probably the least developed of New Zealand's National Parks. In the north-east its beech forests are traversed by State Highway 6, between the Haast and Makarora valleys via the low Haast Pass, and there are good tramping tracks nearby in the Wilkin, Young and Blue valleys; in the south-west it is crossed by the famous Routeburn Walk, a three-day crossing from the head of Lake Wakatipu to the Milford Road via Harris Saddle. There is also a five-day circuit of the Mount Earnslaw massif and the Forbes Range via the Rees and Dart valleys. Virtually everywhere else there is untracked mountainous wilderness: fearsome gorges, idyllic grassy flats, hanging valleys with swift clear rivers, slabby schist peaks, and about 100 glaciers over the length of the park.

One of the most unusual areas is the Red Mountain 'mineral belt' around the Cascade and Pyke valleys in the south-west. Here the screes of rusty-red ultramafic rock are largely devoid of plants except for stunted 'serpentine shrubland', which seems able to tolerate the toxic levels of magnesium in the soils.

The park has the usual diversity of habitats between the wet west and the drier east. The dappled light in large Red Beech trees on the floor of the lower Dart Valley is a visual enchantment. The Red and Silver Beech forests are important habitats for the Rifleman and the Yellowhead (Mohoua), a gregarious warbler sometimes called the Bush Canary. Kea and New Zealand Falcon are found in the park, and the alpine herb-fields are among the best remaining habitats for the Rock Wren.

The Haast River near the Gates of Haast road-bridge, above its confluence with the Landsborough River.

RIGHT At the southern end of Mount Aspiring National Park, the Routeburn Track crosses the Humboldt Mountains to the Hollyford Valley in Fiordland National Park.

BELOW The diminutive Rock Wren (*Xenicus gilviventris*), among the most engaging and tantalizing of New Zealand's birds and the sole survivor of the ancient New Zealand Wren genus *Xenicus* now the Bush Wrens (varieties of *X. longipes*) are considered extinct. It is the only New Zealand bird capable of overwintering in the alpine zone, which it does in a state approaching hibernation.

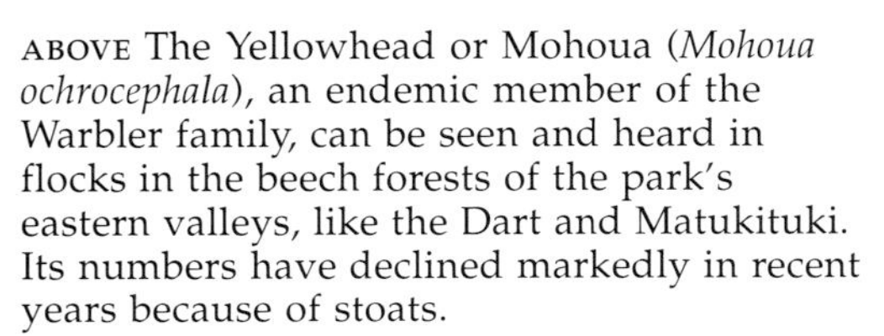

ABOVE The Yellowhead or Mohoua (*Mohoua ochrocephala*), an endemic member of the Warbler family, can be seen and heard in flocks in the beech forests of the park's eastern valleys, like the Dart and Matukituki. Its numbers have declined markedly in recent years because of stoats.

RIGHT The Routeburn Falls, one of several on the lip of the hanging valley of the Routeburn as it begins to plunge steeply into the beech forest.

ABOVE Mount Aspiring (3,030 metres [9,941 feet]) from the west, across the lower Bonar Glacier. It is easy to see why it is sometimes called the 'Matterhorn of the South'. During the last glaciation the three surrounding glaciers – the Volta, Therma and Bonar – honed this isolated peak to its classic horn shape.

LEFT The upper Barrier River lies in a hanging valley near the Olivine Ice Plateau. Many such remote alpine valleys make up the Olivine Wilderness Area, a wild central core of the park.

BELOW LEFT Gnarled Silver Beech or Tawhai (*Nothofagus menziesii*) trunks make up most of the higher-altitude forests in Mount Aspiring National Park's wetter western parts. Here, near the tree-line on the Routeburn Track, the branches are covered in moss and Old Man's Beard lichen (*Usnea* spp).

BELOW *Gaultheria crassa*, an endemic member of the widespread Heath family, likes rocky, low-alpine habitats.

BOTTOM *Raoulia* species, found in fell-fields throughout the Southern Alps, are called 'vegetable sheep' because of their hummocky form and woolly texture.

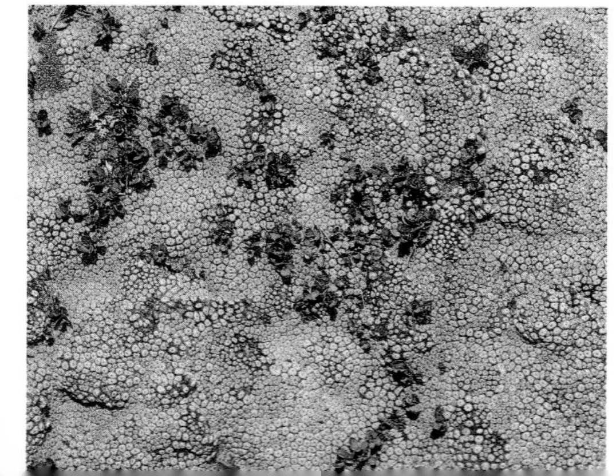

FIORDLAND

Fiordland is a land apart. No other region of New Zealand possesses such a uniformly wild and spectacular character. Most visitors have neither the time nor the skills to penetrate the fastnesses of this remotest corner of the country; they have no need to, for the excitement of driving along the Milford Road, cruising the waters of Milford Sound or walking the Milford or Kepler tracks adequately serves as the experience of a lifetime. For Fiordland is simply overwhelming in the scale of its physical difficulties: everywhere are the sheer rock walls of the mountains, deep fiords penetrating far into the interior, thousands of lakes and waterfalls, and always the rain and the sombre moss-laden Silver Beech forest. Virtually all of the region is uninhabited, protected in perpetuity as Fiordland National Park, one of the world's great wilderness areas. The area of the park is 1,257,000 hectares (4,850 square miles), or five per cent of the area of New Zealand as a whole; only with the designation of Kahurangi National Park in 1994 did it cease to be the case that Fiordland National Park's area was as large as that of all New Zealand's other 11 National Parks put together.

Glaciers remain only in the Darran Mountains and Wick Mountains around Milford Sound, but their phenomenal power in carving the U-shaped valleys and fiords is evident everywhere. Since glacial ice was the sculptor, and tectonic forces the uplifter, of Fiordland's mountains, why is this landscape so different from the rest of the mountainous hull of Te Waka a Maui, where ice and uplift were just as extensive? The main reason is the ability of Fiordland's very hard, crystalline igneous and metamorphic rocks to withstand erosion, thereby largely retaining their morphology since the end of the last glaciation, about 14,000 years ago.

Maori tradition has a very dynamic explanation for the shape of the fiords. Tu te Rakiwhanoa, the son of Aoraki, on finding his father and uncles turned to stone, began the great labour of reshaping the hull of the upturned canoe to make it more suitable for human habitation. He started his work in the south, chopping into the smooth mountain wall. His early work was messy, leaving many islands and ragged coastlines (Preservation Inlet and Dusky Sound). By the time he reached the north he had perfected his technique, leaving the clean-adzed walls of Piopiotahi (Milford Sound) as a monument to his skill. The handiwork of Tu te Rakiwhanoa is reflected in more than 400 Maori place-names throughout Fiordland.

Fiordland has always had a romantic history of human endeavour. The fiords were regularly visited by the Maori because of the rich *kai moana* (seafood) and *pounamu* (greenstone). The prized translucent jade *takiwai*, found near the mouth of Piopiotahi, was particularly sought-after. Subsequently sealers and whalers came, riding the 'Roaring Forties' to shelter in the myriad anchorages of the southern fiords. There were unsustainable efforts at mining and timber extraction; climate and isolation ultimately defeated all these attempts to colonize this wildest of places.

Explorers probed the interior, searching for ways through the maze of mountains; names like Sutherland, Grave, Talbot, Homer and Mackinnon recall attempts to find a safe overland route into the fiords. Their endeavours were ultimately successful, despite the terrible privations they suffered; by 1890 the 'Finest Walk in the World' – the Milford Track – had been opened. Carving out the Milford Road was an even more dangerous enterprise, a 20-year saga of inching forward with pick and shovel in the face of deadly avalanches and rockfalls.

Fiordland was the scene of New Zealand's first truly national conservation victory – the Manapouri Campaign of the 1960s and 1970s. For the first time in New Zealand's history of resource development, the all-powerful hydroelectricity industry was stopped from exploiting to the maximum the waters at its disposal. It was the most important test for the 'preservation in perpetuity' intent of the National Parks Act. As a consequence, the level of Lake Manapouri, the most beautiful of the eastern great lakes in the park, was not raised. The lake remains one of the scenic gems of Fiordland, to be enjoyed while walking the Kepler Track or while crossing to Doubtful Sound.

Fiordland has a wonderful variety of wildlife habitats. There

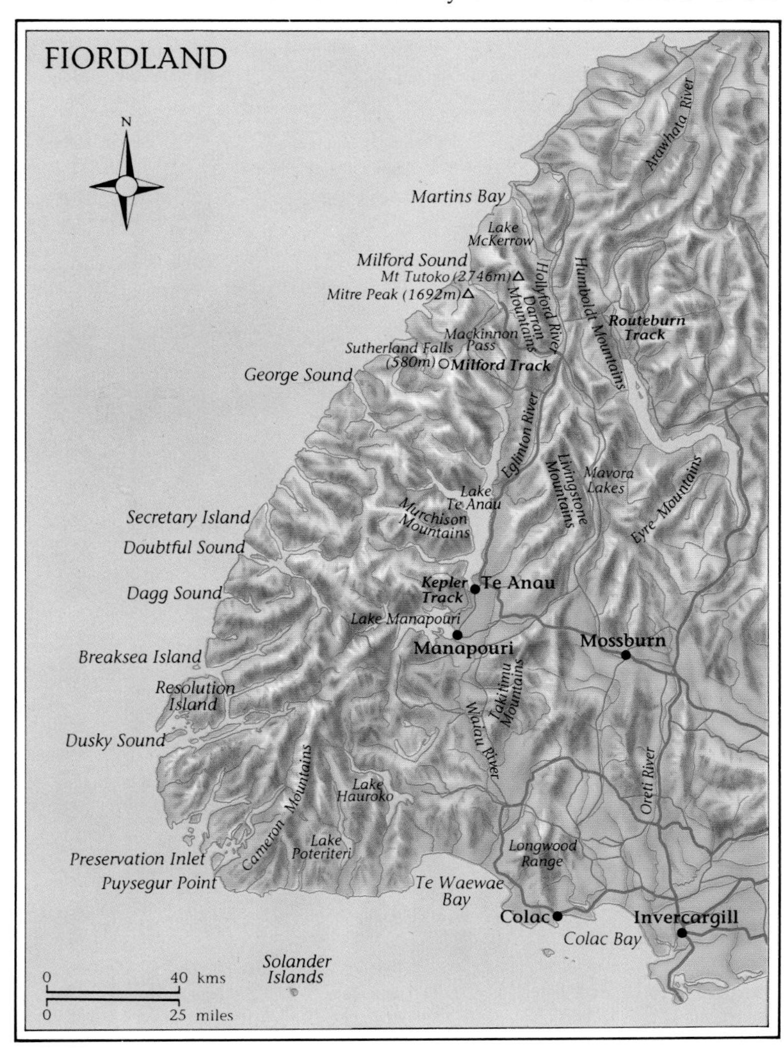

are pristine rocky coastlines and small sandy beaches which are still covered in native Pingao, the golden sand-binding plant. Rich podocarp-beech forests hug the coastal fringes and cover the alluvial flats of the valley floors and the remarkable series of marine terraces of Waitutu, on the south coast. Silver Beech forest, festooned in moss, often grows from sea level to bush-line. The subalpine grasslands that extend the entire length of Fiordland are so diverse that they support nine different species of Snow Tussock (*Chionochloa*). And, everywhere, colourful bog vegetation flourishes around the thousands of small wetlands. Fiordland is one of New Zealand's six mainland centres of plant diversity and high endemism: over 700 species of higher plants and 24 known alpine plants are endemic to the area.

Fiordland's fauna has some very distinctive elements. One is the extraordinary diversity of insects, which seem superbly adapted to the cool, wet environment. There are estimated to be over 3,000 invertebrate species, probably about ten per cent of which are endemic to this region; although Te Namu (the pestilent Sandfly) is the most notorious of these, the 100 or so species of brightly coloured day-flying alpine moths are particularly interesting. The rare Fiordland Skink is another curiosity; predators have eliminated it from all but a few islands in the southern fiords, but there it can be found in writhing profusion in the harsh splashzone of the exposed shoreline rock platforms.

One man, Richard Henry, as 'Custodian and Caretaker of Resolution Island' fought a lonely battle for 15 years (1894–1909) to save the birdlife from stoats and other introduced predators, but his heroic efforts were doomed to failure. Nevertheless, the region's birdlife is not without interest. The Fiordland Crested Penguin is a very handsome inhabitant of the nooks and crannies of the long coastline, and two rare birds are synonymous with the mystique of Fiordland: the Takahe and the Kakapo. The Takahe is a large flightless rail which was thought extinct until being 'rediscovered' in 1948 in a remote alpine valley in the Murchison Mountains. The Kakapo, sadly, is probably now extinct in Fiordland; the last beleaguered survivors were transferred from their lonely Milford valleys to sanctuaries on Maud Island (see page 122) and Little Barrier Island (see page 75) during the mid-1970s.

Lake Te Anau, the eastern gateway to Milford Sound and most of Fiordland National Park, is the largest lake in the South Island, covering more than 350 square kilometres (135 square miles).

Milford Road and Milford Sound

The Milford Road, running 120 kilometres (75 miles) from Te Anau to Milford Sound, traverses a superb cross-section of Fiordland National Park. It is worth beginning the journey at the park's visitor centre, on the shores of Lake Te Anau, where you should first view the displays and audiovisuals and obtain information on the various nature walks and interpretation sites to visit along the road. For the initial 40 kilometres (25 miles) the road follows the eastern shore of the lake and then enters the Eglinton Valley, with its attractive grassy flats and forests of large Red Beech trees. Mirror Lakes provide a popular scenic walking venue close to the road; and the 'Avenue of the Disappearing Mountain', which derives its name from the illusion of mountains sinking into the beech forests lining the road ahead, is another attraction.

The Divide between the Eglinton and Hollyford rivers is the lowest pass – 530 metres (1,738 feet) – on the mountain chain that extends from Nelson Lakes to Fiordland. Here the three-day Routeburn Track begins, but a rewarding half-day trip to the alpine tops of Key Summit, above Lake Howden, is also available. Around the Divide the forest consists of marvellously gnarled Silver Beech trees, and the slopes above abound with U-shaped hanging valleys, marking the path of tributary glaciers that once flowed into the main valley glacier. The upper Hollyford Valley ends abruptly in a cirque wall pierced by the famous Homer Tunnel. This is a forbidding place, especially in early summer when the scars of spring avalanches are still obvious. Near the tunnel portal, the Homer Alpine Walk is a highlight of the journey; Kea entertain the visitor with their antics, and from late November to February there are rock gardens of wild alpine flowers.

The descent through the wild Cleddau Valley includes a side-walk to the Chasm, where the river has cut an impressive slit gorge through a band of diorite rock. A sudden view of glacier-clad Mount Tutoko – at 2,746 metres (9,003 feet) the highest peak in Fiordland – is obtained just before reaching Milford Sound. A cruise on the sound is an unforgettable experience: Bowen and Stirling Falls, Mitre Peak, the Lion, Mount Pembroke and Sheerdown Peak all encircle this most spectacular of fiords.

RIGHT Mitre Peak, the best-known landmark in Milford Sound, reaches only 1,692 metres (5,551 feet) above the water level but looks much higher because its smooth rock walls drop sheer to the perfect 'U-shaped' Sinbad Valley (on the left) and the fiord itself (on the right). The vegetation on these steep fiord walls is easily saturated by the rainfall – up to 10,000 millimetres (394 inches) per annum – and then the entire slope of soil and vegetation, sometimes for more than a kilometre (0.6 miles), may avalanche off into the fiord's deep waters.

LEFT From its Tasman Sea entrance, Milford Sound reaches 13 kilometres (eight miles) into the steepest and highest mountains of Fiordland National Park.

BELOW Bowen Falls empties the Bowen River from its hanging valley directly into Milford Sound. The term 'hanging' denotes the glacial origin of the valley, formed where a smaller tributary glacier once flowed into a main valley glacier. When the glaciers retreated the tributary valley was left 'hanging', often with waterfalls to the main valley.

BELOW RIGHT On the rolling uplands east of Lake Te Anau lie the Red Tussocks (*Chionochloa rubra*) of the Burwood Bush Scientific Reserve, 2,000 hectares (7¾ square miles), now an important site for the captive rearing of Takahe. The reserve is the last protected remnant of the magnificent Red Tussock grassland that once covered the upper Oreti and Mataura catchments and much of the Southland Plains.

ABOVE Heaps of debris from past rock avalanches occur in places around the shores of Milford Sound. Because of the mild temperatures and high rainfall, forest quickly covers this avalanche debris.

BELOW The Eglinton Valley's superb beech forests mix Red, Silver and Mountain Beech species. This even-aged young stand has colonized a low terrace close to the river.

BELOW RIGHT Once thought extinct, Takahe (*Porphyrio mantelli*) was rediscovered in Fiordland in 1948: by 1980 only 120 birds remained, but an intensive captive-breeding programme is showing positive results.

BOTTOM *Aciphylla horrida*, which grows in moist alpine tussock herb-fields, is, like other Speargrasses (Wild Spaniards), renowned for needle-tipped leaves and impressive flower stems.

ABOVE Bottlenose Dolphins (*Tursops truncatus*), a highlight of a cruise on Milford Sound.

RIGHT Part of the road to Milford Sound follows the Hollyford River's upper reaches, climbing through Silver Beech forest.

BELOW The Eglinton Valley is the Milford Road's route through the Livingstone Mountains.

The Milford Track

For more than a century the Milford Track has been known simply (if rather immodestly) as the 'Finest Walk in the World'. The weather can be fickle – it usually changes at least four times during the four-day walk – but the Milford landscape is always spectacular, whether sunlight or waterfalls are playing on the rock walls. The route of the walk, running 54 kilometres (34 miles) from the head of Lake Te Anau to Milford Sound, follows an old Maori trail that once gave access to the *pounamu* of Piopiotahi. In 1888 a track was cleared by the explorer Quintin Mackinnon, and he became the first tourist guide. Conditions were fairly primitive for the parties that began walking the track in 1890, but some hospitality was provided by pioneer hoteliers Donald and Elizabeth Sutherland on the shore of Milford Sound. In the days before the completion of the Homer Tunnel in 1953, walkers also faced an arduous journey out again, via the rugged Esperance River and the snow traverse from the Grave-Talbot Saddle to Talbot's Ladder above Homer Saddle. Today about 10,000 trampers per season walk the track, staying in either private or Department of Conservation hut accommodation.

Highlights of the journey include the Sutherland Falls – three leaps, totalling 580 metres (1,900 feet) – and the many forest contrasts: the more open Red, Silver and Mountain Beech forests of the Clinton Valley, and the moss-covered Kamahi and Silver Beech forests of the much wetter Arthur Valley. Subtle differences in the composition and colour of the forests on the valley walls highlight the paths of huge avalanches of water-saturated vegetation and soil, some up to 800 metres (2,620 feet) long. The summit of Mackinnon Pass, at 1,073 metres (3,518 feet), is the visual climax of the journey. Here the alpine tarns reflect the peaks and the herb-fields are alive with celmisias and gentians, Mountain Buttercups and Snow Berries, while screeching Kea soar on the updrafts, much to the delight of weary travellers. Behind, the impressive U-shaped canyon of the Clinton lies below the ramparts of Castle Mount. Mount Balloon soars directly ahead and, to the north, the eye is captured by the sweep of the Jervois Glacier below Mount Elliot. Waterfalls and occasional avalanches plunge from the glacier into the deep cirque at the head of Roaring Burn, the route that the walker must take to descend into the Arthur Valley far below.

LEFT The Milford Track's first stage follows the Clinton River up through beech and Mountain Ribbonwood glades before climbing to Mackinnon Pass.

ABOVE Mackinnon Pass – at 1,073 metres (3,520 feet) the highest point on the Milford Track – offers a dramatic view back into the classical U-shaped valley of the Clinton River.

LEFT Waterfalls are everywhere along the Milford Track and Milford Road thanks to high rainfall and to hard rocks that are very resistant to water erosion.

BELOW Kea are a popular sight for walkers on the Milford Track, who can watch them dive and wheel in the winds that pour through the Mackinnon Pass.

ABOVE This 'strawberry' fungus is a species of *Cyttaria*, a parasite on Silver Beech. *Cyttaria* infections cause the formation of woody galls on branches and stems.

LEFT Near Quintin Hut on the Milford Track, the Sutherland Falls, reputedly the fifth highest in the world, plunge 580 metres (1,900 feet) in three leaps from their source in Lake Quill.

ABOVE In the south and on some subantarctic islands Prickly Shield Fern (*Polystichum vestitum*), found throughout New Zealand, often forms a near-impenetrable cover on the forest floor.

RIGHT Groves of Mountain Ribbonwood (*Hoheria lyallii*), draped in trailing mosses, occur along the wetter western parts of the Milford Track.

BELOW LEFT Widespread in mountain regions from East Cape southwards, the Everlasting Daisy gains its name from the persistent bracts enclosing the small papery flowers.

BELOW CENTRE Among the most visible alpine herbs is the Mount Cook Buttercup, the world's largest buttercup, with its impressive white flowers from December to January.

BELOW RIGHT *Celmisia semicordata*, the largest New Zealand Celmisia, is widespread in the South Island high country, except in the driest mountains south of Nelson.

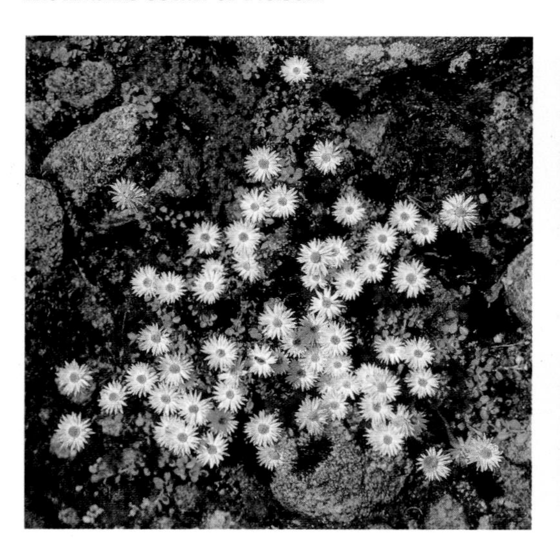

Fiordland from the Sea

South of Milford Sound, the western Fiordland coastline, facing the Tasman Sea, stretches for another 200 kilometres (125 miles) to Puysegur Point. But the true length of this coastline is probably more than 1,000 kilometres (620 miles), because another 14 fiords breach the outer wall. Some of the largest ones, like Doubtful, Breaksea and Dusky sounds, extend back more than 30 kilometres (18.6 miles) into the heart of the Fiordland mountains. South of Dusky Sound the land begins to change as the mountains retreat from the coast. On the south coast, facing remote Solander Island, there has been uplift over the last million years to produce a landform unique in New Zealand: the Waitutu marine terraces, a series of forest-covered former shore platforms (in places, up to 11 can be recognized), with the oldest uplifted to 1,000 metres (3,280 feet) above today's sea level.

The two largest islands in the fiords – Resolution and Secretary islands – failed as wildlife sanctuaries because stoats and deer proved capable of swimming the channels, 1-2 kilometres (0.6-1.2 miles) wide, which separate them from the Fiordland massif. Modern science, however, has succeeded in exterminating Norway Rats from Breaksea Island – 170 hectares (420 acres) – a conservation landmark as significant as the eradication of cats on Little Barrier and possums on Kapiti.

Fiordland's marine environment is also unique. The combination of heavy rainfall, humic staining of the water (producing water which is the colour of tea but otherwise pure) and sheer fiord walls has produced a layer effect. The dark freshwater layer, three metres (10 feet) deep, sits on top of the warmer (by several Celsius degrees) saltwater, and consequently only weak yellow-green light can filter through. This suppresses usual seaweed growth, allowing fragile, many-branched black corals and other normally deepwater species (such as ancient brachiopods and sea-pens) to thrive at shallow depths, only about a tenth of that found elsewhere in the world.

Doubtful Sound is among the longest western fiords – 35 kilometres (22 miles) – and the only one besides Milford Sound to be reachable by road.

ABOVE Dusky Sound, one of the largest of the southern fiords, is famous for its many beautiful forested islands, its superb anchorages and its association with Captain James Cook's first two visits to New Zealand.

BELOW CENTRE The world's largest black coral population grows in the dimness of the fiord walls. Here a Snake Star (*Astrobrachion* spp) has entwined itself around the coral's branches.

BELOW LEFT Sea pens (*Sarcophyllum bollonsi*) are among many unusual marine organisms occupying a unique restricted habitat in relatively shallow water – at 10-40 metres (33-130 feet) – on the fiord walls.

BELOW RIGHT Red brachiopods (or filter-feeding 'lampshells') occur in this same shallow marine zone. Normally known only on the deep sea bed, they are among the most ancient forms of marine life.

The outer headlands of Fiordland's southern fiords are exposed to the full forces of westerly storms, with winds regularly reaching gale force as they surge around the southern Fiordland mountains to pour through Foveaux Strait.

ABOVE The moss-covered Silver Beech forest between Dagg Sound and Crooked Arm, one of several places where the fiords are separated merely by narrow necks of land.

RIGHT Now rare, Pingao (*Desmoschoenus spiralis*) is a colourful native colonizer of sand dunes. Its rhizomes produce cord-like roots that penetrate deep into shifting sands, stabilizing the dunes.

BELOW LEFT The endemic South Island Robin (*Petroica australis australis*) can be a charming companion as it fossicks about on the forest floor for insects and worms.

BELOW RIGHT Breeding only on the remotest shores of the South Island and Stewart Island, the Fiordland Crested Penguin is one of the world's rarest penguins.

OTAGO AND SOUTHLAND

East of the mountains of Mount Aspiring National Park lies the lake district of Otago – great glacial troughs now filled by lakes Hawea, Wanaka and Wakatipu. These lakes, celebrated for their breathtaking scenery (and their trout fishing), provide much of the visual appeal of north-west Otago: the reflections on Lake Hawea from the Haast highway, the golden autumn colours around Lake Hayes and Glendhu Bay on Lake Wanaka, or the winter alpine majesty of the Remarkables rearing up above Queenstown on the shores of Lake Wakatipu.

The lakes are a world away from the overcast and misty Otago coastline (with its prolific marine life), yet there is a link: the magnificent, albeit now largely tamed, Clutha River. The largest river in New Zealand, the Clutha has its source in Lake Wanaka, being thereafter soon joined by the Hawea and then by the equally big Kawerau River, which drains Lake Wakatipu and the Richardson Mountains. Taking 338 kilometres (210 miles) to reach the Otago coast, just north of Nugget Point, the Clutha traverses one of the wildest and emptiest landscapes in the country: Central Otago.

Central Otago is separated from Canterbury by the Hawkdun Range and the Kakanui Mountains, a long arm of greywacke rock and tawny tussock grasslands which extends right down to the coast at Shag Point. The main inland scenic route (State Highway 8) crosses this barrier via Lindis Pass, perhaps the most accessible and magnificent Snow Tussock landscape in the South Island. But Lindis Pass is more than just a provincial marker: it is also a major geological boundary, denoting the transition from the angular greywacke landscape of Canterbury to the shiny, flat, slabby schist landscape of Otago. It is schist that gives Central Otago its distinctive landscape character: the craggy rock pillars (tors), the shiny metallic grey streambeds (from mica in the schist), and the stone huts left over from a bygone era of exploration and hardship – for, to New Zealanders, schist rock is synonymous with the romance of the Otago gold rushes of the 1860s.

The mountains of Central Otago are quite different in shape and colour from the high mountainous backbone in the west. The topography is much more subdued; the summits of the ranges are broad and rolling, the remnants of one of the oldest land surfaces in New Zealand – a peneplain formed during the Cretaceous period. Subsequent parallel faulting of the peneplain generated a very distinctive blocky landscape, a series of basins and ranges oriented all in the same direction. So regular is the pattern of these block mountains that Central Otago is often referred to as basin-and-range country.

LEFT The wild Shotover river, flowing from the heart of the Richardson Mountains to join the Kawerau River just below the outlet of Lake Wakatipu, is steeped in Otago's gold-mining history.

In addition to its distinctive landforms, Central Otago has the most unusual climate in the country. Completely landlocked by encircling mountains, it is screened from the moist westerly and south-easterly winds. This rain-shadow effect is so pronounced that 'Central' has the most 'continental' climate in all New Zealand, with extremes of searing summer heat, dryness and winter cold. This semi-arid climate is most pronounced in the Clutha and Manuherekia valleys around Alexandra – with only 350 millimetres (13.8 inches) annual rainfall – but the Maniototo Basin and the Taieri Valley near Middlemarch are further dry pockets. In these locations a number of inland salt lakes and saltpans can still be found – some of them possibly relics from ancient inland seas – but they are sorely in need of special protection: they contain some of the country's rarest plants and insects.

Except for some small shelf glaciers on the summits of the Garvie, Pisa and Old Man ranges, Central Otago was not glaciated during the ice ages, and consequently these mountains do not show the gouging, stripping and rejuvenating effects of the glaciers that carved Fiordland and all the lands flanking the Southern Alps. Nevertheless, the higher areas – above 600 metres (1,970 feet) – were subjected, as they still are, to extremely severe climatic conditions: freezing temperatures, high winds and freeze-thaw cycles which greatly reduce soil structural stability, and even sort schist fragments into regular patterns called 'stone polygons' and 'stone drains'. The summits of the block mountains are a wonderland of periglacial phenomena, notably soil-stripes (where differential freezing of the soil has sorted it into crests and

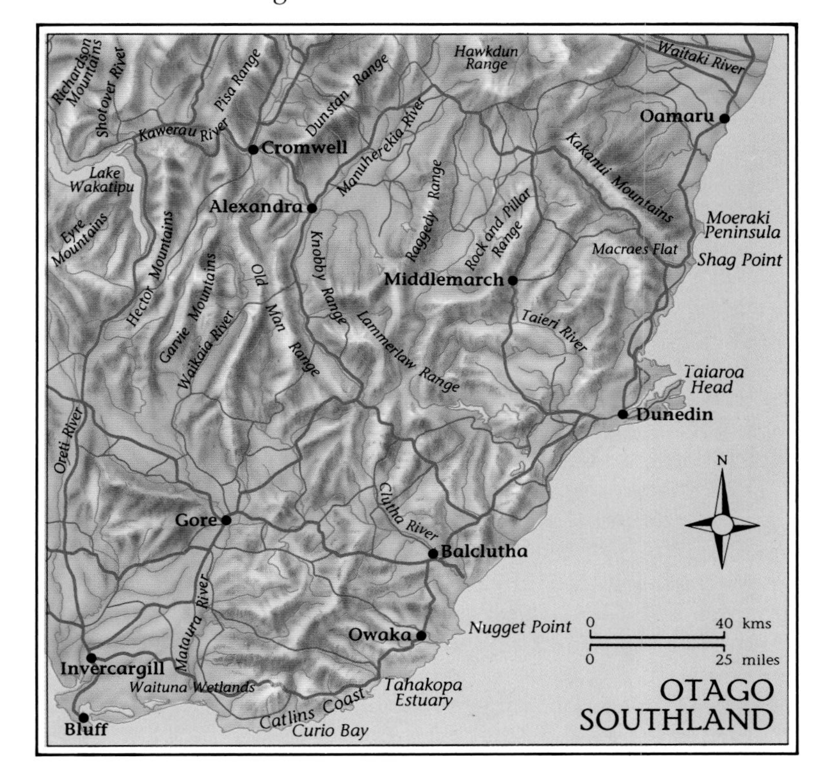

furrows), hummocks and solifluction lobes (where soil has 'flowed' like treacle) – all features rarely found in the much steeper and wetter Southern Alps.

Podocarp forests, mainly the hardy Hall's Totara, Bog Pine and Celery Pine, became established in the cooler, moister localities of this semi-arid climate after the end of the last glaciation. Subsequently, natural fires caused a partial replacement by beech forest. Then centuries of fires associated with Maori moa-hunting favoured the spread of tussock grasses (species of *Festuca*, *Poa* and *Chionochloa*), inducing an almost prairie-like appearance, especially in the uplands around the Lammerlaw, Lammermoor and Knobby ranges. Today the tussock grasslands have been degraded by fire, rabbits and stock overgrazing, but the landscape still retains much of its indigenous character through the survival of hardy plants like Matagouri, Speargrass, Scabweed, Kowhai and Coprosma. Indeed, the unique combination of climatic extremes, old and relatively stable landforms and the isolation of the block-mountain 'islands' has made Central Otago another mainland centre of endemism. The region is one of the highest priorities for conservation because of the small number of existing reserves and threats to the survival of its many distinctive plants and animals.

The coastlines of both Otago and Southland are noted for their varied scenery and prolific wildlife. This south-east corner of the South Island has a number of geological and oceanic features which together give it a very distinctive maritime environment. Unlike the erodible coasts along the Canterbury Plains, the south-east coasts have several hard-rock localities which shield the land from the powerful ocean currents surging in around Fiordland from the southern Tasman Sea. Notable examples are the Otago Peninsula (an ancient volcano now breached by the sea) and the Moeraki Peninsula (with a popular geological curiosity, the Moeraki Boulders, lying on the beach just to the north); and south of the mouth of the Clutha River lies another, Nugget Point, where the Murihiku Escarpment (marking the true geological boundary between Otago and Southland) plunges into the ocean. Further south stretch the magnificent rocky headlands and bays of the wild and remote Catlins coastline, each marking the coastal end of the great parallel folds of the South Otago Uplands. Southland, too, has its hard-rock 'anchors': the small volcanic remnant of Bluff Hill and the boulder-strewn seascapes around Colac Bay, where the volcanic rocks of the Longwood Range terminate in the shallow, stormy waters of Foveaux Strait.

In contrast to the deep waters around the Fiordland coast, the continental shelf extends for over 30 kilometres (19 miles) off the Otago coast and 100 kilometres (62 miles) to the south of Stewart Island. Shallower than 150 metres (490 feet), these waters support a wide variety of marine mammals. Hector's Dolphin and Hooker's Sea Lion are noteworthy around Moeraki and Shag Point, while the Otago Peninsula and Catlins coast have an intricate network of estuaries and headlands with many haul-out and breeding areas for a variety of seals, including New Zealand Fur Seal, Southern Elephant Seal, Leopard Seal and Hooker's Sea Lion. Further south, Orca (Killer Whale) and Right, Sperm and Pilot Whales move through Foveaux Strait.

The variety of seabirds, many quite rare or endangered, is a major attraction of this coastline. The most imperilled is the endemic Hoiho, or Yellow-eyed Penguin. Other penguins include the endearing little Southern Blue Penguin and the occasional straggler from the several species of Crested Penguin that live around the islands of the Southern Ocean. A notable coastal wildlife feature is the Royal Albatross colony at Taiaroa Head – the only mainland breeding colony of albatross in the world. The shallow estuaries and lagoons around Invercargill (New River, Bluff Harbour and Awarua Bay) rank with Lake Ellesmere and Farewell Spit as the most important wading-bird habitats in the South Island. Nearby lies one of the botanically most interesting parts of the Southland coast, the cushion bog of Waituna Lagoon in Toetoes Bay. Here the cold climate and peaty soils have induced a unique moorland, a sea-level community containing plants which usually occur only in alpine regions.

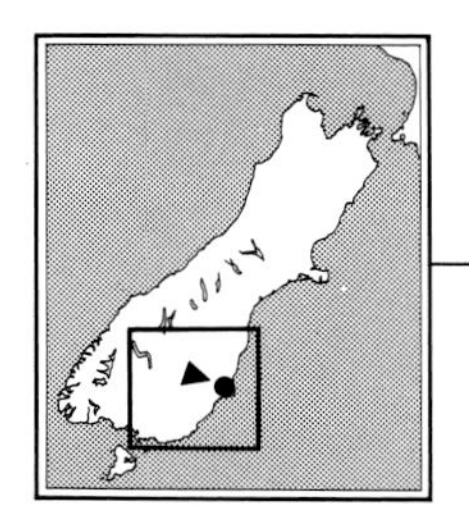

Otago Peninsula

Dunedin relishes its reputation as the 'wildlife city of the south'. It is a maritime city at the landward end of Otago Harbour, a reach of the sea 20 kilometres (12½ miles) into the heart of the ancient volcano which we now know as Otago Peninsula. This is a place of many coastal moods: the cold Southland current sweeps past the peninsula, sea fogs cling to the smooth moor-like slopes and cliffs, and winter south-easterly storms often bring snow to high points like Flagstaff, Swampy Summit and Mount Cargill. Much of the distinctive character of the peninsula stems from its geology, particularly the soaring columns of basalt in attractions like the Organ Pipes (on the slopes of Mount Cargill), the cliffs of Lovers' Leap, and Blackhead on the south coast. The columns huddle tightly together above a jumbled apron of collapsed rock, each eroded block a perfect prismatic fragment of the parent column; in places the waves have sorted the basalt blocks from these cliffs to form spectacular black-boulder beaches, like that south of St Clair.

There are two wildlife highlights for visitors to the peninsula. Each involves a carefully guided seasonal walk, one to observe the breeding colony of Royal Albatross at Taiaroa Head at the entrance to the harbour, the other to glimpse the return at dusk of the world's rarest penguin: the Hoiho, or Yellow-eyed Penguin. The albatross symbolizes all that is wild and free in the vast Southern Ocean, yet about 75 years ago it chose to begin a colony on the outskirts of a major city – the only known example of such behaviour. The story is a saga of painstaking research and predator control by dedicated individuals in a long, often frustrating effort to protect these remarkable birds. Visitors can expect to witness about 20 pairs of birds at any one time during most of their protracted 11-month breeding cycle.

Hoopers and Papanui inlets are the two main estuaries on the Otago Peninsula's south-eastern side. Both have important saltmarsh communities.

LEFT A Jewelled Gecko (*Heteropholis gemmeus*) perching on a *Coprosma propinqua* shrub on Otago Peninsula. One of New Zealand's most attractive geckos, with its brilliant colouring and diamond-shaped markings, the Jewelled Gecko shows more variation in colour than any other of its genus.

BELOW LEFT Of the world's 33 species of shags (cormorants), 13 breed in the New Zealand region. The cliff-nesting Spotted Shag (*Stictocarbo punctatus*) is among the most abundant, breeding mainly on the east coast of the South Island. Freshwater shags can be identified by their black feet while marine shags usually have pink feet – but the Spotted Shag, although it fits into the 'marine' category, has orange feet, unlike any other New Zealand Shag.

BELOW The Yellow-eyed Penguin (*Megadyptes antipodes*) is one of New Zealand's best-known penguins, thanks to a highly publicized conservation programme, but remains very rare: continued threats include predation (often by dogs), habitat loss, set nets and fluctuations in food supply. There are breeding colonies around the Otago Peninsula, right on Dunedin's doorstep: tourists come to watch the birds dash up the beach to their nests beneath coastal flax and shrubs.

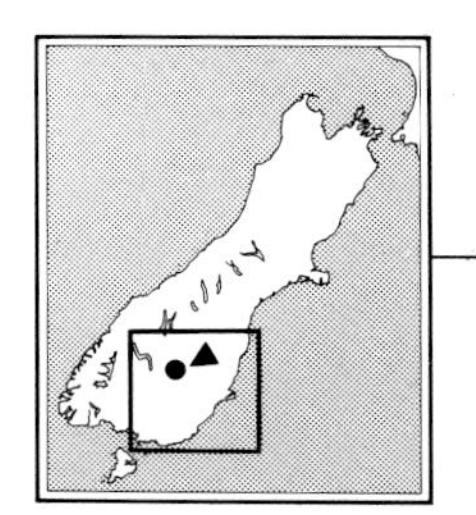

Central Otago Block Mountains

The names of some of the block mountains of Central Otago evoke their craggy, wild character: Rock and Pillar Range, Raggedy Range, Rough Ridge and Knobby Range. Landmark schist tors, like Leaning Rock on the Dunstan Range or the Obelisk on the Old Man Range, have been likened to the Leaning Tower of Pisa. The Pisa Range itself is in the west, along with the Remarkables and the Hector Mountains; these peaks are taller – 1,900–2,300 metres (6,230–7,540 feet) – and more accessible because of their proximity to highways; further east, at altitudes of 1,600–1,900 metres (5,250–6,230 feet), lie the remote alpine wildernesses of the Garvie, Old Woman and Old Man ranges. The grasslands of Slim Snow Tussock, Blue Tussock and Alpine Fescue Tussock on the slopes of these block mountains give way on the summits to a tundra-like sward of cushion-forming plants. This community is dominated by the dwarf shrub *Dracophyllum muscoides* (the smallest member of this widespread subalpine genus) and cushion mats of *Raoulia hectori*, *Phyllachne rubra* and *Hectorella caespitosa*. This intriguing dwarf vegetation preserves the shape of the soil-stripes and hummocks from the erosive power of the savage winds. Above the remote Nevis Valley, on the western side of the Garvie Mountains, lies an internationally important low-alpine wetland containing string bogs and island-studded tarns.

Two groups of animals are noteworthy in these inhospitable mountain environments. The very rich and interesting insect fauna (at least 1,200 species, many endemic to these mountains) includes beetles, grasshoppers, wetas, cicadas and moths. Most are large and colourful, like the race of black- and yellow-banded Mountain Weta (*Hemideina maori*) found on the summit of the Rock and Pillar Range.

The other animal group comprises the rare and vulnerable 'giant' skinks of Otago. These two species – Otago Skink and Grand Skink – have a curious distribution, restricted to the Lindis Pass area and, far to the east, the Middlemarch Uplands. Both are considered among New Zealand's largest and most beautiful skinks, although their habitats have been greatly reduced by 120 years of pastoralism and predation.

The summits of the Dunstan Range typify Central Otago's block mountains: schist tors protrude through thin soils and scattered cushion plants.

ABOVE The Milkweed genus *Euphorbia* has about 2,000 species worldwide but only one native to New Zealand – the Shore Spurge (*E. glauca*), found on the coastline of the Otago Peninsula but rare elsewhere in the country.

TOP LEFT Macraes Flat, near Middlemarch, one of the few known habitats of Otago's rare 'giant skinks' (Otago Skink and Grand Skink). They survive among the schist tors and tussocks of gorges where farming is impossible.

CENTRE LEFT The Otago Skink (*Leiolopisma otagense*) grows to 25 centimetres (ten inches) long. Although its range is greater than once thought, it is nevertheless restricted to a few inland Otago sites; even there, continuing land degradation is threatening this susceptible species.

BELOW LEFT Climate is severe in the Rock and Pillar Range: a late-summer snowfall on the summits highlights the soil-stripes and hummocky ground. These broad summits are the habitat of a striking black-and-yellow-banded race of Mountain Weta (*Hemideina maori*).

BELOW *Pterostylis mutica*, a Greenhooded Orchid, flowers in spring, producing several flowers on each stem.

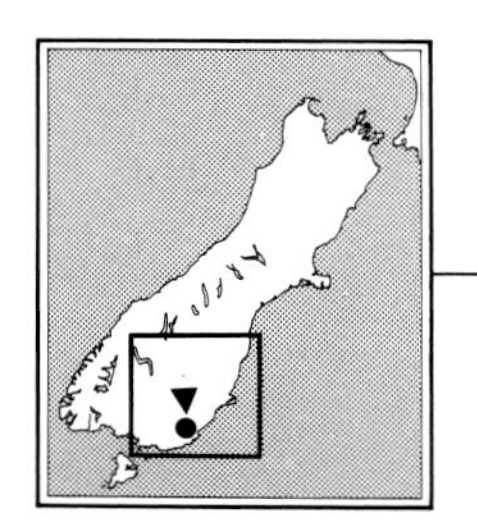

The Catlins Coast and Forests

The Catlins coastline is the only part of the east coast of the South Island where the original forest still extends down to sea level. It is a wonderful, wild place, far from busy highways and tourist centres. The Catlins Forest Park comprises 60,000 hectares (232 square miles) of tangled wilderness, made up of Rata, Kamahi and podocarp forest; the ranges are relatively low – up to 700 metres (2,295 feet) – but the small rivers draining them form an intricate trellis-like pattern, flowing through deep gorges before emerging in peaceful estuaries like those of the Catlins, Tahakopa and Waikawa rivers.

The coast offers an outstanding scenic and ecological experience. The sedimentary rocks have been carved into a range of features: there are cliffs up to 200 metres (655 feet) high and a wide range of caves, blowholes, arches, rocky islets and headlands. Yet between the headlands lie many natural beaches. Three of these – Tahakopa, Tautuku and Waipati – are the outstanding natural beaches of the eastern South Island. Each has intact foredunes covered in Pingao and other native sand-binding plants. In the older dunes at Tautuku nestles little Lake Wilkie, its surrounding forest ablaze at Christmas with the profuse crimson flowers of the Rata.

At the northern end of this coastline lies spectacular Nugget Point, now both a Scientific Reserve and a Marine Reserve. It is the only New Zealand location where the Elephant Seal, New Zealand Fur Seal and Hooker's Sea Lion coexist, has one of the largest breeding colonies of Hoiho (Yellow-eyed Penguin), is a feeding-ground for just about every species of seabird found on the Otago coast, and harbours many unusual plants, invertebrates and fossils.

At the southern end of the Catlins coast lies Curio Bay, where the fossil remains of an ancient forest, dating from the Jurassic period, can be reached on a rock platform at low tide.

The Clutha enters the ocean at Molyneux Bay. The rocky, indented Catlins coast, running 80 kilometres (50 miles) southwards, teems with wildlife.

LEFT At Curio Bay on the southern Catlins coast are fossils of a Jurassic forest. Yellow-eyed and Fiordland Crested Penguins can often be seen here in late summer.

ABOVE Curio Bay's silica-impregnated Jurassic treestumps and logs resemble modern Kauri, Norfolk Pine and Matai. The forest was probably preserved long ago in a flood of volcanic mud and debris that then hardened into rock under the sea.

BELOW LEFT On the South Island's eastern coast, Nugget Point is outstanding for geology and wildlife. In this area, uniquely, New Zealand Fur Seal (*Arctocephalus forsteri*), Elephant Seal and Hooker's Sea Lion coexist.

BELOW In early summer New Zealand Fur Seals come ashore to breeding colonies like this one at Nugget Point.

STEWART ISLAND

Stewart Island is, at 172,200 hectares (665 square miles), the smallest and southernmost of the three main islands of New Zealand. In Maori legend, the island is linked with Te Waka a Maui (the 'Canoe of Maui', or South Island); its oldest name was Te Punga o te Waka a Maui (the 'Anchor of Maui's Canoe'). Indeed, the two islands were connected during the Pleistocene ice age, for Foveaux Strait, whose waters are even today quite shallow, was then dry land. Geologically, the island is related to Fiordland and Southland, with their igneous rocks. Its northern third consists of diorite highlands, around Mount Anglem – the island's highest point, at 980 metres (3,213 feet) – and the peaks and spires of the Ruggedy Mountains; this region is traversed by the North-West Circuit Track, offering ten days of enthralling wild coasts and bottomless peat – enough to satisfy the most enthusiastic of backpackers! A faulted depression running through Paterson Inlet and the swampy floor of the Freshwater River separates these northern mountains from the southern two-thirds of the island, where lie Mount Rakeahua and the wilderness of the remote granite uplands of the Tin Range, plus the Deceit Peaks and Fraser Peaks far to the south, around Port Pegasus.

Superficially, the whole island looks as if it was extensively glaciated during the ice age – not by the deep U-shaped-valley glaciers of Fiordland but by extensive ice-sheets that smoothed and rounded the contours of the ranges. But this was not in fact the case: there was no glaciation except for small local cirque glaciers around the high points of Mount Anglem and Mount Allen (in the Tin Range). Instead, the spectacular rock landforms of the mountains are the result of powerful physical weathering of the hard crystalline rocks when exposed to the relatively dry but very cold and windy conditions prevailing in this area during the final stages of the last ice age, when it had become too cold for forest to survive. In such conditions granite will flake off exposed surfaces to leave sheer monoliths or domes quite devoid of vegetation – something like the well known landscapes of Rio de Janeiro or California's Yosemite Valley. The most spectacular examples on Stewart Island lie far to the south, in the wilderness around Port Pegasus, especially the Granite Knobs, in the head of the Robertson River, and Gog and Magog, in the Fraser Peaks. These striking landforms were largely unknown to New Zealanders until 1977, when the rare bird the Kakapo was 'rediscovered' in this most inhospitable of habitats.

Stewart Island is, quite simply, the best remnant of wild New Zealand the average visitor who seeks an insight into what the pre-human environment must have looked and sounded like is ever likely to experience. The singular charm of Stewart Island lies in its combination of the natural and the spectacular, extending right across its mountains, forests, coastlines, islands and marine environments. It is still almost completely clothed in indigenous vegetation, with, along the coasts, the dense podocarp-hardwood forest extending right to, or even overhanging, the water's edge. Hundreds of islands and rocky islets ring the coastline, particularly around Paterson Inlet and the intricate eastern and southern sides of the island, where the sea has flooded back into the narrow river valleys as the land has been tilted to the east.

The climate is cloudy, windy and wet, and the soils predominantly poorly drained peats, defying all past attempts to establish traditional farming. There are only about 500 permanent inhabitants, who live around Oban in Half Moon Bay, their livelihood increasingly based on sustainable fishing and tourism. Most of the island is protected as conservation land, much of it in the form of strict nature reserves. Nevertheless, there is a long tradition of use by the Kai Tahu, especially the annual harvesting of 'muttonbirds' (Sooty Shearwater) from the Titi (Muttonbird) Islands off the island's north-east and south-west coasts.

The forests of Stewart Island differ in a number of ways from their nearest neighbours in Waitutu (see page 154) and the Longwoods, and, for that matter, from the forests of the subantarctic islands to the south. They are true rainforests, the southernmost podocarp forests in the country and probably the world. When the climate began to warm again at the end of the last glaciation, about 14,000 years ago, birds began to carry the seeds of podocarps and hardwoods like Kamahi back to the island, across the exposed land where Foveaux Strait now is, from forest refuges on the mainland; Southern Rata was able to re-establish itself probably because the light seeds could travel easily on the winds. But beech, mainly dependent on dispersal in streamwater, did not cross Foveaux Strait before the sea flooded back. It is not the only common tree absent: others include Celery Pine, Mountain Cedar and Kowhai. Most forest trees are Rimu, Miro, Kamahi and Southern Rata; other podocarps, Kahikatea, Matai and Hall's Totara are locally important, and Yellow-Silver Pine becomes more predominant in the poorly drained montane areas, especially in the south. Introduced browsing mammals have

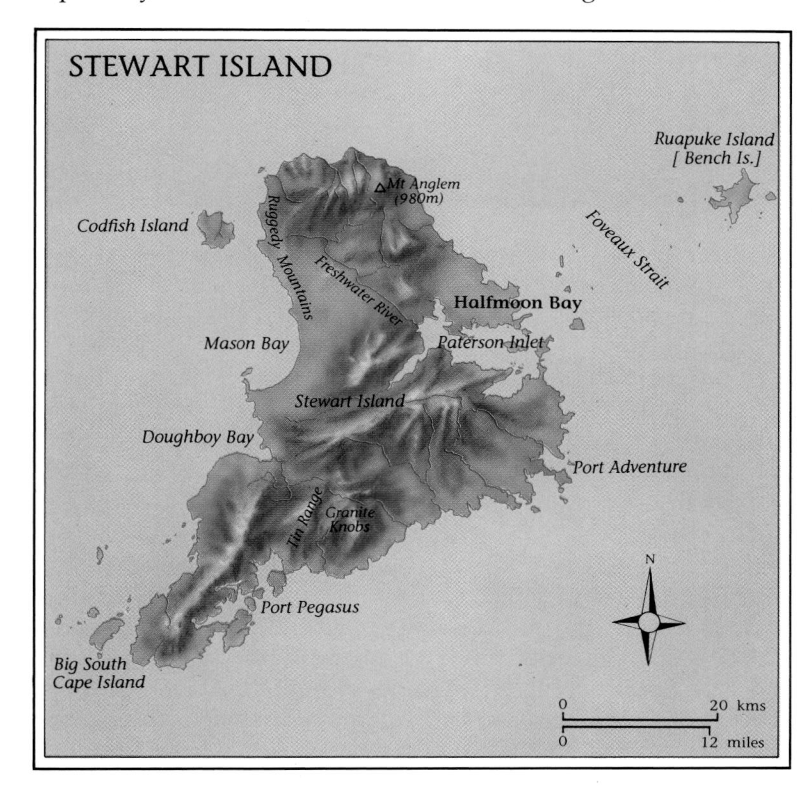

significantly modified these forests, however: possums, Red Deer and Virginian Deer were introduced at the turn of the century and have subsequently eliminated many of the tastier broad-leaved shrubs.

In the more exposed coastal areas the forest is replaced by a tight hedge of wind- and salt-resistant shrubs, especially the tree daisies like *Olearia* and *Brachyglottis* (referred to locally as 'muttonbird scrub'). A large section of the Stewart Island coastline, around the north-west from Smoky Beach to Mason Bay, is particularly exposed to the extraordinarily strong westerly winds coming through Foveaux Strait. Here lie the wildest sandy beaches and dune systems in New Zealand, and their names capture some of the violence of the onshore winds, constantly driving salt spray and sand further inland: Smoky, East and West Ruggedy, and Big and Little Hellfire. Dune systems running parallel with the prevailing north-west airstream can be traced up and over 'sandpasses' behind the Ruggedy beaches, past the Ruggedy Mountains, and far down the wetlands of the mid-Freshwater River – in all, as far as 20 kilometres (12½ miles) from the coast, a distance matched only by the dune systems of the Manawatu on the west coast of the North Island. Further south lies Mason Bay, an arc 15 kilometres (9½ miles) long of sandy beach backed by some of the highest dunes in New Zealand, such as the Big Sandhill, which stands 156 metres (511 feet) tall above the Mason Bay hut. To sit on the Big Sandhill and watch the sun set across Mason Bay, while the wind whips the sand and the dune grasses below into a whirling, shimmering frenzy of motion and light, is to really feel the awesome power at work in this island wilderness. And yet native sand-dune plants – golden Pingao, the grass *Austrofestuca littoralis*, Sand Pimelea and Silver Tussock – do a remarkable job of battling these winds and binding the sand together. Gradually it develops a topsoil and is colonized by larger plants like Flax, *Coprosma propinqua*, 'muttonbird scrub' and, eventually, Rata-Kamahi forest. These sand-dune communities are

OPPOSITE Paterson Inlet, a large drowned river-valley system in eastern Stewart Island, is the only accessible, rock-walled inlet in New Zealand to remain completely surrounded by natural vegetation. Seen distantly is Ulva Island.

RIGHT Forest – typically an association of Rata, Rimu, Kamahi and Miro (never Beech) – covers most of Stewart Island.

BELOW Easily recognizable with its beadlike threads, Neptune's Necklace (*Hormosira banksii*), the only species in its genus, is a brown alga common on rocky platforms around New Zealand and Australia.

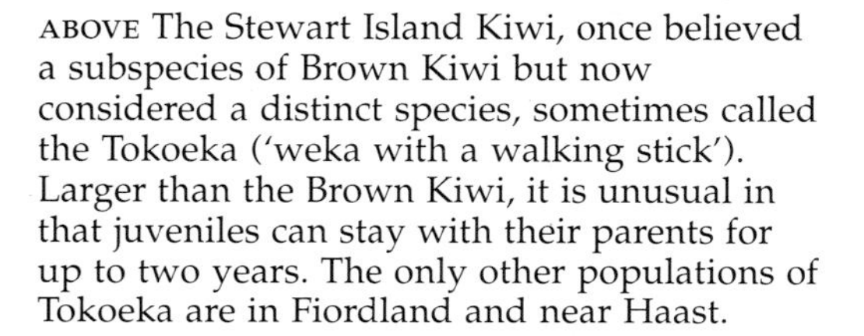

ABOVE The Stewart Island Kiwi, once believed a subspecies of Brown Kiwi but now considered a distinct species, sometimes called the Tokoeka ('weka with a walking stick'). Larger than the Brown Kiwi, it is unusual in that juveniles can stay with their parents for up to two years. The only other populations of Tokoeka are in Fiordland and near Haast.

RIGHT Many small streams meander through the sandhills of Mason Bay, a 15-kilometre (10-mile) windswept sandy arc on the island's west and the habitat of the rare dune plant *Gunnera hamiltonii*.

ABOVE LEFT A Variable Oystercatcher (*Haematopus unicolor*) and chick in dunes at Mason Bay. This bird, endemic to New Zealand, has three colour phases, but only the black is found on Stewart Island.

ABOVE The Kakapo (*Strigops habroptilus*), sole member of its endemic genus, is one of the world's most endangered birds as well as the world's largest parrot and the only flightless one. Only 60 birds remain, all on New Zealand's offshore islands.

LEFT The rare Harlequin Gecko (*Hoplodactylus rakiurae*), a hardy resident among subalpine cushion plants of southern Stewart Island, is among the world's southernmost lizards.

the most unmodified in the country, along with those of Te Paki in the far north (see page 56) and Fiordland (see page 153).

The wildlife of Stewart Island is both diverse and prolific, despite the ravages perpetrated by introduced rats and cats on some of the indigenous fauna. Fortunately mustelids have not reached the island, and this may account for some differences with Fiordland. The first thing to strike the visitor is the constant sound of the forest birds – by contrast with so much of New Zealand's forest, which sadly has now fallen silent save for the sighing of the wind. Here there are the chorus of forest song and the exuberant cry of flocks in the canopy or passing overhead – Tuis and Bellbirds, parakeets and pigeons, Kaka and cuckoos, fantails and Grey Warblers, Tomtits and Brown Creepers. The Stewart Island Kiwi and subspecies of Robin, Weka and Fernbird can be seen, or at least heard. The Stewart Island Kiwis of the Mason Bay dune tussocklands and shrublands are cherished by visitors to this extraordinary place; these bold creatures, unlike other kiwis, are not afraid to leave their burrows to feed in daylight.

The most famous forest birds on Stewart Island were the last wild population of Kakapo, 'rediscovered' in 1977 in the southern wilderness around Port Pegasus. Initial studies within their extensive track-and-bowl systems showed that the birds were under serious threat from feral cats. The numbers of this endearing but so very vulnerable 'old New Zealander' dropped alarmingly during the mid-1980s, and, in a track-and-capture campaign of heroic proportions, the last 40 or so were transferred to rat- and cat-free island sanctuaries – Maud, Codfish, Little Barrier and Mana islands (see pages 75, 122 and 181). Today there are about 25 birds on Codfish Island, just over three kilometres (1.9 miles) off the coast opposite the Ruggedy Range. Although the island has Kiore, it is free of Ship Rats and Norway Rats.

Other birds are still having to fight for survival on the main island. In one of the most severe locations on Stewart Island, high on the northern end of the Tin Range around Table Hill, there is a small, beleaguered population of an endemic wader, the New Zealand Dotterel. The only other breeding-grounds for this coastal bird lie more than 1,000 kilometres (620 miles) away, along the North Island coastline between Ninety Mile Beach and East Cape. The plight of this threatened colony, probably now numbering no

RIGHT Nearly four kilometres (2½ miles) of turbulent ocean separates Codfish Island, a nature reserve and one of the Kakapo's last sanctuaries, from the distantly seen Ruggedy Mountains of Stewart Island's northeast coast.

BELOW Paterson Inlet's shallow, sediment-free humus-stained waters are an important habitat for a group of the most ancient marine lifeforms: the filter-feeding 'lampshells' (or brachiopods). The fluted red *Terebratella sanguinea* is one of several free-living species on the inlet's floor.

RIGHT Bare granite knobs, a landform unique in New Zealand, are a striking feature of the upland wilderness of southern Stewart Island; they occur through successive layers of rock flaking off. This locality in the southeast Tin Range was one of the last remaining Kakapo habitats on the Stewart Island mainland.

more than 20 breeding pairs because of cat predation, underscores the extreme difficulty of the task facing the Department of Conservation, WWF and other wildlife conservators in saving vulnerable animals on the main islands.

Stewart Island also provides the most graphic and best documented example of the vulnerability of New Zealand's fauna to introduced predators – even on supposedly safe islands. In about 1962 Ship Rats managed to reach Big South Cape Island off the south-west coast of Stewart Island. By early 1964 they had reached plague proportions, completely overrunning the island and ruining much of its wildlife. The losses were enormous: the Stewart Island Snipe and Bush Wren (both now extinct) were wiped out, as were the Stewart Island Robin and Fernbird; and the South Island Saddleback, Greater Short-tailed Bat, parakeets and bellbirds were all but extinguished. In a last-ditch rescue operation, the Wildlife Service managed to capture the survivors of the rare South Island Saddleback and transfer them to some small rat-free islets in the Muttonbird Islands.

On a happier note, the first marine reserve on Stewart Island is proposed for Paterson Inlet. This inlet, with its 190 kilometres (120 miles) of forested coastline and 20 small islands, is one of the largest and least-modified estuaries in the country. Most of the 380 types of seaweed which occur around Stewart Island can be found in this relatively shallow – 15–25 metres (49–82 feet) – inlet. Its waters are sheltered and sediment-free, although stained with tannins from the island's peat soils. These conditions have some similarities with those in the fiords (see page 164), and are ideal for the survival of one of the most ancient forms of marine life: brachiopods (or filter-feeding 'lampshells'), which date back more than 500 million years in the fossil record. The brachiopods of Paterson Inlet are of great scientific interest in that they include free-living species, such as the pink-coloured *Neothyris lenticularis*. In addition, the shallow bed of the inlet abounds in scallops, oysters, anemones, tube-worms, sea cucumbers and the red seaweed *Rhodymenia*; closer to shore, the rocky shallows are covered in kina (sea eggs), limpets, chiton and paua (abalone). Swirling forests of the Great Kelps *Macrocystis* and *Durvillea* trail from the rocky shores in the outer reaches of the inlet, providing a haven for reef fish like Blue Cod and Moki.

THE SUBANTARCTIC ISLANDS

The wildest fragments of New Zealand are the subantarctic islands lying hundreds of kilometres to the south and east of Stewart Island. Of the five groups, the Auckland Islands are by far the largest, at 62,560 hectares (242 square miles); Campbell Island, at 11,330 hectares (44 square miles), and the Antipodes Islands, at 2,100 hectares (8.1 square miles), are quite significant in size; while the Snares, with 328 hectares (1.3 square miles), are the closest to the mainland of New Zealand, and the remote Bounty Islands, with 135 hectares (334 acres), are mere specks in the vast Southern Ocean. They are all true oceanic islands, none of them having been connected to the New Zealand mainland during the last ice age. What they lack in size, however, they more than make up for in biological importance. Along with Australia's Macquarie Island, they are the only land in the Pacific sector of the Southern Ocean. In consequence they are of international conservation significance as breeding-grounds and haul-out havens for the millions of seabirds and thousands of marine mammals that inhabit the cold waters around Antarctica.

Strictly speaking, the islands are cool-temperate in environment because they lie north of the Antarctic Convergence; in other words, they have a mean annual air temperature above 5 degrees Celsius (41 degrees Fahrenheit) sufficient to allow the growth of some trees and other woody plants. Nevertheless, it is common in New Zealand to refer to them as the subantarctic islands.

The islands' climate is indescribably bleak. They lie in the 'Roaring Forties' and 'Furious Fifties', latitudes swept incessantly by strong wet westerly winds. They are also high islands, the main ridge-crests of Campbell Island and the Auckland Islands lying between 400 metres and 650 metres (1,300–2,130 feet) above sea level. While their annual rainfall, at 1,200–1,800 millimetres (41–71 inches), is not high compared with much of mainland New Zealand, it is remarkably constant, with precipitation on more than 300 days of the year. Consequently sunshine is very limited: Campbell Island, for example, has only 660 hours of sunshine annually, a mere one-third of that in Christchurch. But the winds are probably the most oppressive feature of the islands, for they seem never to stop; over the entire year the winds scouring Campbell Island, to use it again as our example, have a mean hourly speed of more than 30 kilometres (19 miles) per hour. Such winds were sought out by nineteenth-century sailing ships as they plied these latitudes from west to east, and today's round-the-world yacht racers likewise strike far into southern waters to capitalize on the speed of these winds.

The Auckland Islands in particular were, and still are, a major obstacle for such mariners. The history of the Auckland Islands is an epic saga of seal slaughter and failed settlement, shipwreck and privation, and the bravery of castaways in their desperate attempts to survive. The island groups are still unable to support any permanent human population except the Department of Conservation station maintained on Campbell Island for weather forecasting and scientific research.

The five island groups all lie on the wide Campbell Plateau, a submarine platform with relatively shallow waters, up to 1,000 metres (3,280 feet) deep. Their geological make-up varies, however. The Snares and Bounties are composed of granite, reflecting the underlying continental nature of the Campbell Plateau. The Auckland Islands, Antipodes Islands and Campbell Island are very old volcanoes, with many impressive landforms,

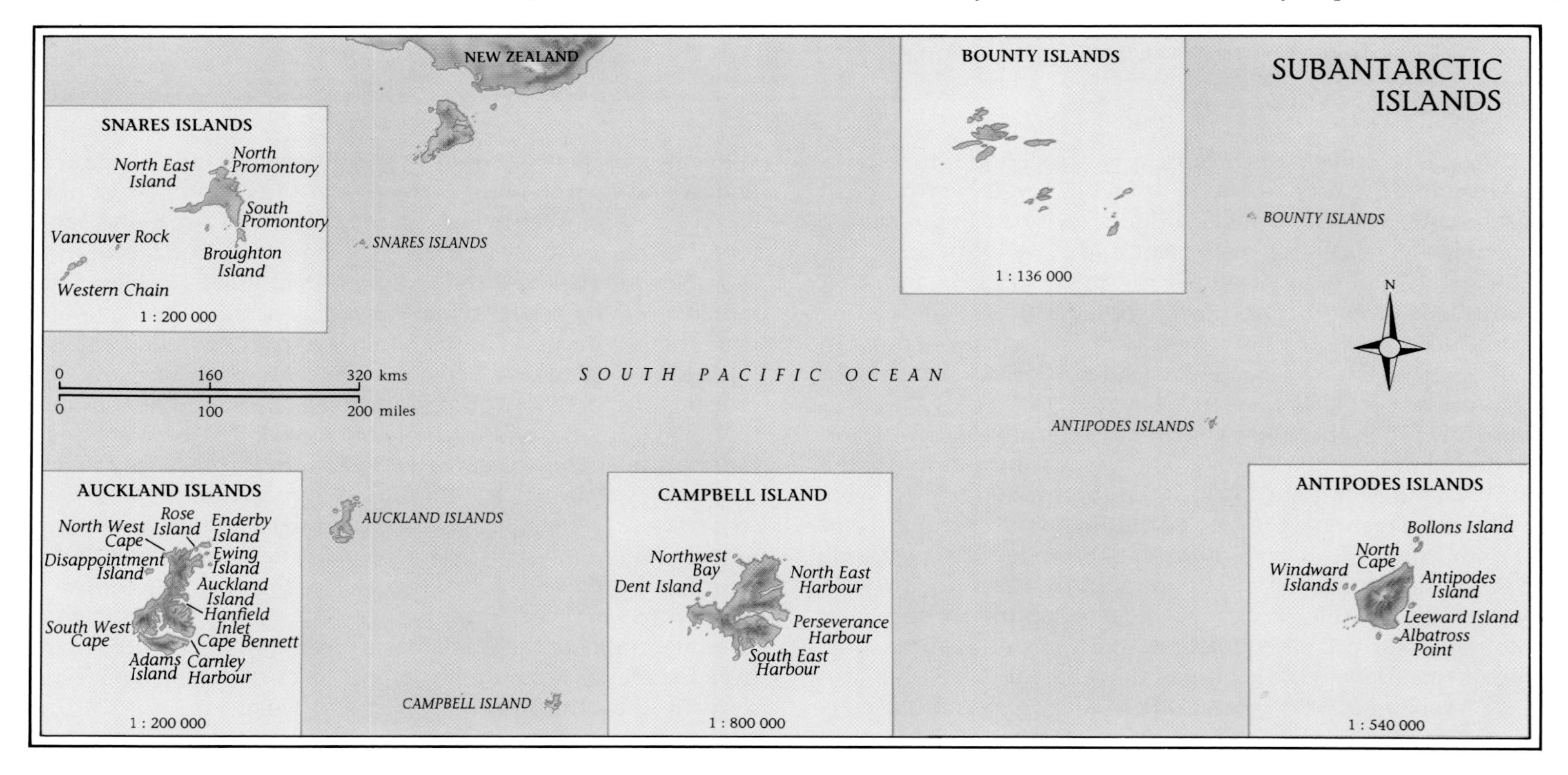

The view across Northwest Bay on Campbell Island shows limestone cliffs streaked with veins of volcanic lava. Fine days like this are rare in the 'Furious Fifties' latitudes of New Zealand's subantarctic islands.

including the caldera of Carnley Harbour, the spectacular cliffs of Adams Island and Campbell Island, and the many erosion-resistant volcanic stacks, like the Windward Islands off the western coast of the Antipodes. Because the islands are so widely separated and so variable in size and geological structure, their marine environments differ in the range of seaweeds and fish. Unfortunately, these environments have not yet been well studied, but what does stand out is the very low diversity of marine plants and fish compared with Otago and Southland, or even Stewart Island. On the other hand, while the species may be limited in number, those found are present in great abundance. This is an enigma which scientists find very interesting but difficult to explain. There are no common brown seaweeds or popular shellfood like paua (abalone), crayfish and scallops, or common edible fish like Tarakihi and Blue Cod. However, one marine plant feature is shared with all the other coasts of southern New Zealand: the fields of the massive Bull Kelp, its huge rubbery limbs holding fast on even the most wave-tossed headlands.

The international conservation significance of these subantarctic islands stems from the evolution of their biotas in isolation over many millions of years. In this sense they are a microcosm of the rest of New Zealand, but far less spoiled: some of the islands are the most pristine in the world. In particular, the Snares Islands, plus Adams and Disappointment islands in the Aucklands group, are among the last islands anywhere to contain floral and faunal communities completely unmodified by humans or introduced animals.

The vegetation of the islands is subalpine in character, and very different from New Zealand's mainland flora. It has universally developed on deep, peaty soils which are acidic and, except where burrowing seabirds have aerated and enriched the soil, of low fertility. The climate and the wet, peaty soils have turned the rolling summits of the islands into a moorland of tussocks, their shaggy manes tossing constantly in the wind. On the Auckland Islands the tussocklands consist of *Chionochloa antarctica*, a species from the same genus as the mainland Snow Tussocks; on the Snares and Antipodes the tussocklands are a mixture of *Poa* species

Within the grasslands there are many of the large 'megaherbs' which bring so much colour and character to the islands. Many belong to familiar genera, like the *Anisotome* (Carrot), *Celmisia* (Daisy), *Gentiana* (Gentian) and *Myosotis* (Forget-me-not), but here they are often endemic species with large and/or highly coloured leaves and flowers. These herb-fields offer an amazing sight to the visitor, even to people familiar with the profusion of the mainland's alpine fell-fields in flower. In the Southern Alps and Fiordland the flowers are generally white, pale mauve or pale yellow, and most of the plants have been restricted to inaccessible ledges because they are so palatable to the introduced deer, Chamois and Tahr. But here in the mist-drenched wild gardens of the subantarctic the yellows are more vibrant; there is the surprise of the blue-flowered *Myosotis antarctica*; a delightful range of pink, mauve, red, deep brown and purple abound in the spectacular species of *Pleurophyllum* (an endemic genus of megaherb, in the Daisy family) and in other plants like *Anisotome antipoda*, *Gentiana concinna* and *Damnamenia vernicosa*. Some of the megaherbs – the several species of *Stilbocarpa*, for example – are attractive for their large glossy leaves as much as for their yellow and purple flowers.

Campbell Island and the Auckland Islands have some of the southernmost forests in the world, but there are none of the tall conifers and beech trees of the mainland. In the sheltered coastal parts of the Auckland Islands are dwarf forests of Southern Rata and the tree daisy *Olearia lyallii*. The forest canopy is tight against the shearing winds and the tan-coloured trunks are gnarled into a fantastic variety of forms, while the peaty ground surface is largely free of shrubs. On Campbell Island the Rata gives up the struggle in favour of hardy *Dracophyllum* and *Coprosma* species, which are still able to form a 'dwarf subalpine forest' five metres (16 feet) tall.

Overall, the diversity of plant-life is roughly proportional to the size of the different island groups: the Auckland group has 196 native taxa, Campbell Island 128, the Antipodes group 68 and the much smaller Snares group only 20, while the tiny Bounty Islands, although of significance for breeding seals and seabirds, are botanically little more than a group of lichen-covered rock pinnacles. The degree of endemism of the subantarctic islands' flora is significant, but not as great as on the Chatham Islands (see page 197).

Perhaps surprisingly, the islands have no reptiles or frogs. Not at all surprising is the fact that their wildlife is dominated by seabirds, although the limited landbird fauna is interesting as well because of its high level of endemism. The extraordinary density of seabirds that can exist in the absence of predatory land mammals may be seen on the small, pristine Snares Islands. Here, of the 23 different species of breeding seabirds, the most populous are Sooty Shearwaters, nearly three million pairs of which have honeycombed the peaty soils with their burrows. They and another resident, the endemic Snares Crested Penguin, are so numerous that they have turned the forest and tussocklands into a busy, squarking and rather rich-smelling 'petropolis', undermining or killing whole groves of trees and tussocks before their colony moves on to another choice piece of real estate. At dawn the Shearwaters take off like a black swarm of locusts from any vantage point; the excitement of their mass return at dusk is considered, by the few who have been privileged to witness it, one of the most enthralling of all wildlife experiences. It is certainly no overstatement to claim in general that the Snares are one of the world's priceless wildlife habitats, to be protected at all costs from invasion by rats and other alien mammals. Perhaps their significance is best illustrated by a simple (and perhaps simplified) comparison: the number of seabirds living on the tiny area of the Snares is roughly equal to the entire seabird population of Great Britain and Ireland.

Some of the other subantarctic islands are likewise fortunate enough still to be free of introduced mammals. Sealers devastated the seal population of the Bounty Islands during the nineteenth century, but fortunately did not introduce any land mammals (or, for that matter, higher plants) into the midst of the teeming mass of mollymawks, shags and several million Erect-crested Penguins that, in consequence, still occupy every rock surface – or so it seems – on these very inaccessible islands.

Disappointment Island, and especially the much larger – 10,100 hectares (39 square miles) – Adams Island, both in the Auckland group, are the other outstanding pristine islands, with Adams perhaps the largest of this kind in the world. The Antipodes group also has very high seabird populations, because the only invader is the relatively benign (if not to invertebrates!) House Mouse. On the other hand, Auckland Island has pigs, cats and mice, although fortunately no rats; Campbell Island has cats and Norway Rats. One of the really exciting challenges facing the Department of Conservation and associated volunteers is the eradication, one day, of many of these alien mammals: the successes of eradication campaigns conducted on Mana (mice), Little Barrier (cats) and Breaksea Island (rats) all show that it can be done, albeit with enormous effort.

Despite the mammalian predators, the most diverse bird fauna is that of the Auckland Islands. There are 17 different breeding species of albatross and petrel, plus four endemic landbirds (a rail, a tomtit, a pipit and a snipe) and the endemic Auckland Island Teal. Campbell Island, too, has an endemic teal, the Campbell Island Teal, but it is confined to tiny Dent Island, offshore; it is probably the rarest teal in the world. While cats and Norway Rats have all but eliminated its burrowing petrels, Campbell Island is the outstanding habitat for albatrosses (five species), especially the Southern Royal Albatross. In all, 14 endemic landbirds are known within these subantarctic islands; however, all except one of them (the Antipodes Island Parakeet) are considered subspecies derived from landbirds blown to the islands from Stewart Island or Australia. This process continues; for example, fantails have successfully colonized the Snares within the last 15 years, presumably from Stewart Island.

No account of the subantarctic islands would be complete without discussion of their marine mammals. Four species of seal breed on, or regularly visit, the islands: New Zealand Fur Seal, Hooker's Sea Lion, Leopard Seal and Southern Elephant Seal. Hooker's Sea Lion is the rarest of the world's five sea lion species. The loss of about 100 individuals of this species annually through drowning in the nets of foreign squid trawlers working the seas around the Auckland Islands has led to an international outcry and pressure on the New Zealand Government to designate a marine mammal sanctuary to protect these endangered animals.

The subantarctic islands also lie on the north-south migratory routes of many Pacific Ocean whales. The sheltered inlets and harbours are very important for the survival of one of the baleen whales, the Southern Right Whale. The whales bear their calves and suckle them in these locations, in the painfully slow restoration of a species that was all but exterminated by whalers in the mid-nineteenth century: two centuries ago there were some 60,000 of them in the Southern Ocean; today there are only about 500.

The conservation significance of New Zealand's subantarctic islands is so high that each of the five island groups has been protected in the strictest category of protected area: nature reserve. Appropriately, Adams Island was the first to be so protected, in 1910; the rest of the Auckland Islands followed in 1934; Campbell

Island came 20 years later in 1954, and the Snares, Bounties and Antipodes in 1961. In 1977 all five groups were declared a National Reserve, giving them the same status as a National Park. In recent years there have been suggestions that their internationally important conservation values be recognized through designation as a World Heritage Area.

Victoria Passage, the turbulent western entrance to Carnley Harbour, an ancient breached volcanic caldera that separates Adams Island from the main Auckland Island.

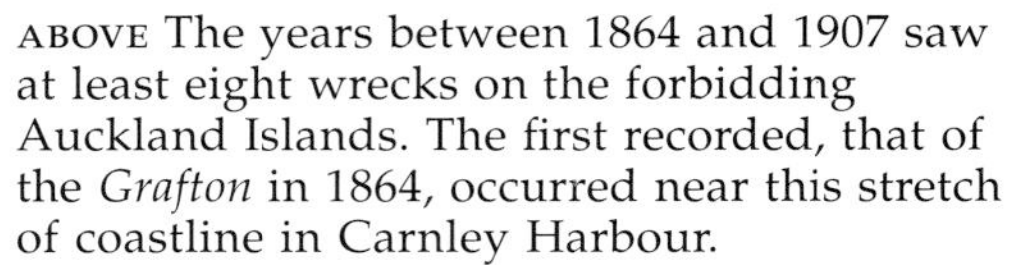

ABOVE The years between 1864 and 1907 saw at least eight wrecks on the forbidding Auckland Islands. The first recorded, that of the *Grafton* in 1864, occurred near this stretch of coastline in Carnley Harbour.

BELOW LEFT A waterfall in Hanfield Inlet, in the Auckland Islands, cascades down lava cliffs. The Aucklands originated mainly from two basaltic volcanoes centred on today's Disappointment Island and Carnley Harbour.

ABOVE A Hooker's Sea Lion (*Phocarctos hookeri*) bull keeps a watchful eye on his 'harem' on the shores of Enderby Island. The dark brown, heavily maned males weigh up to 400 kilograms (880 pounds), their powerful frames equipping them for territorial fights. The graceful fawn females usually weigh only up to 160 kilograms (350 pounds).

LEFT The Giant Tree Daisy (*Olearia lyallii*), a dominant component of forests in the Snares and Auckland Islands. Wind and salt severely affect it, leading to stunted, contorted shapes reminiscent of the mainland's subalpine 'leatherwood scrub'.

RIGHT Hooker's Sea Lion is one of the world's rarest sea lions, numbering about 14,000, with its main breeding population centred on the Auckland Islands. As this female on Enderby Island shows, it can walk on all four flippers, unlike Elephant and Leopard seals, which use a caterpillar-like shuffle.

ABOVE The subspecies of Banded Dotterel (*Charadrius bicinctus*) found on the Auckland Islands is much like its mainland counterpart but slightly larger and more robust; it is estimated that about 730 birds remain.

BELOW *Hebe elliptica*, found in coastal areas from Taranaki south to the subantarctic islands, as well as in South America. Although *Hebe* species occur also in Australia and New Guinea, most are restricted to New Zealand.

ABOVE A Shy Mollymawk (*Diomedea cauta*), the largest of all mollymawks, nests on Auckland Islands cliffs among plants of the colourful megaherb *Anisotome latifolia*.

BELOW LEFT Southern Great Skua (*Stercorarius skua lonnbergi*), a predatory seabird that kills chicks and smaller seabirds, takes eggs and scavenges from large carcases.

BELOW The small, flightless Auckland Island Teal (*Anas aucklandica aucklandica*), a subspecies of Brown Teal. About 1,000 remain, all on outlying islands of the Aucklands.

From above Northwest Bay on Campbell Island, the island's western cliffs present a bold face to the incessant westerly gales. The vegetation here consists of tussocks of *Chionochloa antarctica* and clumps of the striking *Anisotome latifolia*.

ABOVE In September and October Southern Elephant Seal (*Mirounga leonina*) females can be found lolling in the sand dunes and mud hollows of the shores of Campbell Island.

BELOW LEFT Southern Elephant Seal bulls can weigh almost four tonnes and grow to nearly six metres (20 feet) in length, making them the largest flippered marine mammal.

BELOW Hooker's Sea Lion bulls usually arrive at their breeding areas several weeks before the females in order to establish dominance and territory through fighting.

ABOVE Endemic to New Zealand, Southern Royal Albatross (*Diomedea epomophora epomophora*), the world's largest seabird, feeds over southern oceans, coming to land only to nest.

BELOW LEFT *Lycopodium australianum* on Campbell Island, one of only three clubmosses to reach the subalpine zone. Unlike its near relatives it lacks cones and can reproduce asexually.

BELOW The large leaves and brightly coloured flowers of *Pleurophyllum speciosum*, one of the Daisy family, make it the most spectacular of Campbell Island's megaherbs.

ABOVE *Bulbinella rossii*, particularly abundant in the herb-fields of Auckland and Campbell islands because its unpalatable foliage protects it from introduced grazers.

LEFT The tall subantarctic tussock *Poa litorosa*, found from sea level to alpine tops on Campbell Island, survived the island's sheep-farming era when other tussocks were replaced by grassy meadow and herbs were confined to inaccessible cliffs.

BELOW LEFT *Anisotome latifolia*, a giant herb member of the Carrot family, can be a major attraction of the Auckland Islands' tussock grasslands.

BELOW CENTRE A stunted forest, dominated by *Dracophyllum* and *Coprosma* species, grows in sheltered sites on Campbell Island. *Dracophyllum longifolium*, shown here, is not confined to the subantarctic, being typically present in low-fertility leached-soil areas like South Westland's bogs.

BELOW *Stilbocarpa polaris*, reaching two metres (6½ feet) across, typifies subantarctic megaherbs. Sealers and whalers used to eat its corrugated leaves to prevent scurvy.

Snares Crested Penguins (*Eudyptes robustus*) breed only on the Snares Islands in the subantarctic. This bare granite slab, Penguin Slope, is a popular landing spot for the penguins.

The Antipodes Islands, eroded remnants of ancient volcanoes, have spectacular coastal landforms: stacks like the Windward Islands rise sheer from the sea. The high, rolling summits of the main Antipodes Island carry no forest but a wind-tossed sea of *Poa litorosa* tussocks.

ABOVE The volcanic cliffs of the Antipodes Islands provide little opportunity for safe anchorage or access: this inaccessibility is a main protection for these largely pristine ecosystems.

RIGHT New Zealand Fur Seal has slowly recovered since near-extinction in the subantarctic islands during the early nineteenth century.

BELOW Northern Giant Petrels (*Macronectes halli*) can be seen around mainland New Zealand but always return to the subantarctic islands to breed, occasionally in colonies with the closely related Southern Giant Petrel (*Macronectes giganteus*).

ABOVE On the Bounty Islands the gregarious Erect-crested Penguin (*Eudyptes sclateri*) nests on the barren stacks and islets alongside the elegant Salvin's Mollymawk.

BELOW LEFT The Bounty Islands, where Salvin's Mollymawk (*Diomedea cauta salvini*) breed, lack vegetation. This remarkable nest is a pillar of mud reinforced by pigeon feathers.

BELOW The little-studied Erect-crested Penguin, now confined to the Bounty and Antipodes islands, nests just above the shoreline in large, noisy colonies.

THE CHATHAM ISLANDS

The Chathams are quite unlike the rest of New Zealand. This remote group lies 850 kilometres (530 miles) east of Christchurch. The total area is about 97,500 hectares (375 square miles). There are four main islands – Chatham, with 90,650 hectares (350 square miles), Pitt, with 6,326 hectares (24 square miles), South East (Rangatira), with 218 hectares (540 acres), and Mangere, with 113 hectares (280 acres) – plus numerous islets and stacks, many of which are important habitats for seabirds. The climate is depressing: constant winds with cloud cover or low sunshine, moderate temperatures and high humidity. The population of nearly 800 is New Zealand's most isolated, and it is often said that to the islanders the mainland is just as foreign as Australia. They are a frontier people with a history of close dependence on the sea and the land for their survival. The first human inhabitants were Polynesians, who arrived on the islands – which they called Rekohu, meaning 'misty skies' – perhaps 700 years ago, becoming the Moriori people. The peace-loving Moriori developed a distinctive culture, seemingly in balance with nature, but it came to a tragic and abrupt end after their first contact with Europeans in 1791, followed by invasion and enslavement by Maori tribes in 1835. The usual catalogue of resource-exploitation and misery for the indigenous people followed: sealing, whaling, disease, lawlessness and steady clearance of forests and wetlands for the gradual establishment of sheep farms.

Biogeographically the Chathams have an affinity with mainland New Zealand, but with very pronounced changes in their plants and animals because of their long isolation. Of all New Zealand's outlying islands, they show the greatest degree of endemism in their flora and fauna. Of the higher plants, 11 per cent are endemic to the Chathams, including species of *Astelia*, *Aciphylla*, Shrub Daisy, Forget-me-not (the highly attractive blue-flowered megaherb *Myosotidium hortensia*, the only species in this endemic genus), Hebe, Sow Thistle, sedges and Nikau Palm – all of which are threatened or endangered. In addition, 11 of the 12 main forest trees and tree-like shrubs are endemic. Another notable plant is *Sporodanthus traversii*, a distinctive rush-like herb growing up to three metres (10 feet) tall; it is one of the few members of the Restiad family of plants in New Zealand, and, outside the Waikato Basin and Hauraki Plains of the North Island, is found only on deep peats in the Chathams. The major continuing threats to these plants are feral pigs, wandering stock (most of the forest pockets and wetlands are unfenced), possums and the ever-present wind, which carries salt that can, once the protecting shrub layer is broken, desiccate a forest.

Fossil remains show that, prior to human occupation, there were 21 species of petrel, making these islands one of the most prolific breeding-grounds for petrels in the world; eight of these species became extinct during Moriori settlement. There is no fossil evidence of either Moa or Kiwi on the islands, but other species – all extinct by the time Europeans arrived – included a swan, a duck, a sea eagle, a snipe, a coot and a Kaka. Subsequently the Chatham Islands Bellbird, Fernbird, two species of rail, Brown Teal, bittern, New Zealand Falcon and Shoveler Duck have become extinct on the islands. Even today, 42 per cent of the indigenous bird taxa of the islands are endemic, and most of these are threatened or endangered. The only known reptile on the islands is the Chatham Islands Skink, likewise endangered.

The Chathams' marine environment is distinctive and of great economic significance. The convergence of the warm subtropical and cold subantarctic currents over the Chatham Rise leads to a highly productive marine food chain; this is reflected in abundant fish stocks and numerous coastal or open-sea birds. The marine fauna is not well known yet, but is probably a unique assemblage, quite different from central New Zealand's but with similar elements to those of northern and southern New Zealand. Inshore fisheries are a main source of income, with crayfish (lobster), paua (abalone) and Blue Cod being the main commercial species. There are no marine reserves in the Chatham Islands.

Chatham Island itself has a varied topography. Most of it is of low relief, although a series of old volcanic cones stand out in the north, along with many lakes and Te Whanga Lagoon, which – at about 18,000 hectares (70 square miles) – is the dominant landscape feature on the island. The undulating southern tablelands are likewise of volcanic origin, but are now deeply covered with peat. This is the wildest part of the island, rising to almost 300 metres (980 feet) before terminating in an impressive band of cliffs, 200 metres (650 feet) tall and extending 30 kilometres (19 miles) around the entire southern end of the tablelands. The largest remaining forests survive in the Tuku and Awatotara catchments in the south-

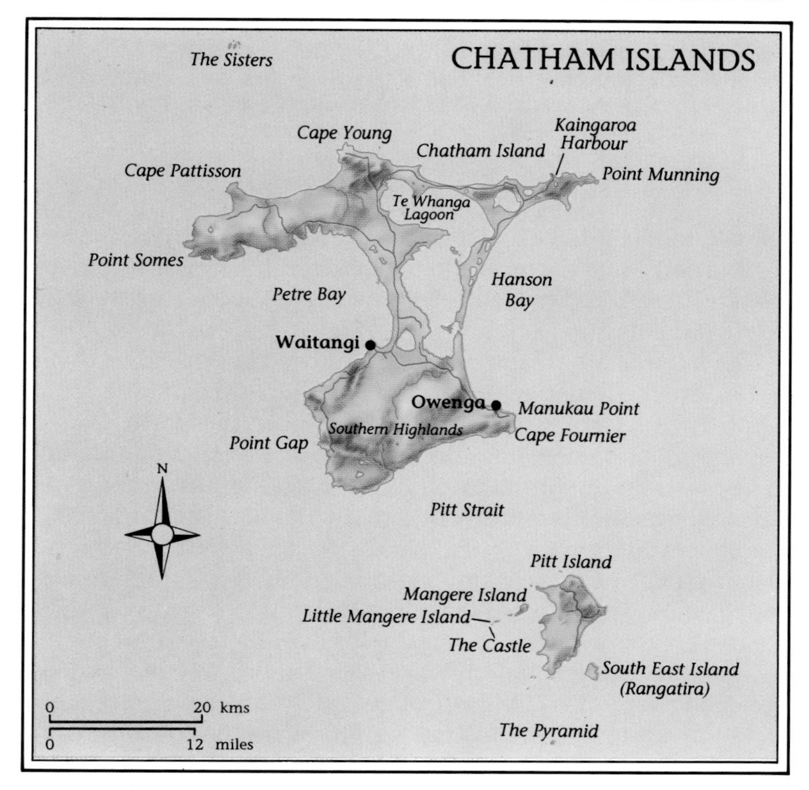

The rugged sanctuary of South East Island (Rangatira), seen from the remote south east coastline of Pitt Island. Rangatira is famous as the home of the Chatham Island Black Robin and of one of the world's rarest wading birds, the New Zealand Shore Plover.

west sector of the tablelands. This is a habitat of critical importance for two of the world's most endangered birds: the Parea (Chatham Islands Pigeon) and the Taiko (Magenta Petrel), each with a known population of under 100 birds.

The latter is a bird shrouded in mystery. It was first recorded in 1867 by the Italian research ship *Magenta* in the middle of the South Pacific Ocean. For many years it was considered extinct. Then, in 1978, its breeding-grounds were discovered on the cliffs of these remote southern tablelands, the result of years of painstaking searches and lonely vigils by David Crockett and other volunteers. Since then, 60 Taiko have been banded by dedicated Department of Conservation researchers; because the numbers are so small, and because the bird returns to its burrow at night, the researchers may capture only a dozen or so Taiko for banding per season. There is still much to learn about this bird: a lone ocean wanderer, it is known to range widely through the Southern Ocean at up to 100 kilometres (62 miles) per hour, continually swooping and diving in its search for squid and other food.

Pitt Island has the Chathams' last areas of original rocky coast vegetation as well as some of the group's best mature broadleaf forest. The island is free of rats and possums, and the 50 islanders are working with conservation management to remove cats and relocate feral sheep. Pitt Island – specifically the as yet unofficial Pitt Island Scenic Reserve – could then provide a haven for endangered Chatham Islands fauna. Although inhabited, the island is still the second largest rat-free island in the Southern Hemisphere; its potential as a wildlife sanctuary – indeed, as a sanctuary that could involve this most isolated of New Zealand's human communities – is one of the most exciting conservation possibilities for the island group.

Mangere and Rangatira are very inaccessible and ringed with sheer cliffs. Both are nature reserves, free from mammalian predators, and consequently support dense breeding colonies of seabirds. Mangere is the main habitat for the endangered Chatham Island Yellow-crowned Parakeet (also known as Forbes' Parakeet) and the rare Dieffenbach's Spaniard, *Aciphylla dieffenbachii*.

Mangere also played a key part in the saving of the endemic Chatham Island Black Robin, and is still one of only two habitats for what used to be the rarest bird in the world. The story of its rescue is now world-renowned. Incredibly, a population of 20–30 Black Robins had managed to survive in about six hectares (15 acres) of scrub forest on the summit of tiny, sheer-cliffed Little

Akeake (*Olearia traversii*) is endemic to the Chatham Islands but has become popular on the main islands as a shelter belt. It is very hardy, growing in unmixed stands on dunes and elsewhere as part of lowland forest associations.

Chatham Island's southern tablelands are the wildest part of the island. An undulating mosaic of forest, *Olearia*/Koromiko/ *Dracophyllum* shrublands and open swampy moorlands, these are the last habitat of the Parea (Chatham Island Pigeon) and Taiko (Magenta Petrel).

Mangere Island (Tapuaenuku) for nearly a century after predators had eliminated them from the rest of the Chathams. Storms and, possibly, the clearing of vegetation for a helicopter landing-pad led to the rapid deterioration of their habitat in the early 1970s. By 1980, there were only five birds left, including just one effective breeding pair: 'Old Blue' and her mate 'Old Yellow'. In the following decade a bold and remarkably successful programme of intensive management in the wild was undertaken by a team led by Don Merton, resulting in a spectacular recovery to 116 birds by 1990. The key to success was his hunch that pairs of robins could be induced to double their egg production if the first clutch was transferred to the nests of temporary foster-parents like the Chatham Island Tomtit. The other crucial move was to transfer the birds to Mangere Island, and then gradually on to Rangatira Island, which has 20 times the area of suitable habitat. 'Old Blue' herself, the matriarch of the Black Robins, was a remarkable bird, living for 13 years – twice the usual Black Robin lifespan – and ensuring the survival of her species. It is entirely appropriate that she is firmly enshrined in New Zealand's wildlife folklore, with her name and picture gracing the national awards given by the Royal Forest and Bird Protection Society to the most deserving of workers for conservation.

Rangatira Island ranks as one of New Zealand's most important island sanctuaries for reasons beyond merely its role in ensuring the survival of the Black Robin: it is the main habitat for the Chatham Island subspecies of Snipe, Tomtit, Tui and Red-crowned Parakeet, and is the only breeding site for the rare Chatham Islands Petrel. The most endangered resident of Rangatira is the New Zealand Shore Plover, now one of the world's rarest wading birds. Early last century this Plover was widespread throughout all of New Zealand's rocky coastlines and estuaries, but it is particularly susceptible to predation by Ship Rats and cats, and by the 1880s it was extinct on the mainland, although a small population (today numbering only about 120) survived on Rangatira. It is a priority species for captive breeding at the Mount Bruce National Wildlife Centre, in the lower North Island, and recent successes with the programme there mean that a second population will probably be established on a predator-free island closer to the mainland.

The Chathams story, then, is a very mixed one: some outstanding modern triumphs beginning to brighten the dark historical picture of appalling loss of habitat and biota – sadly, the Chathams represent probably the worst example of human impact causing loss of indigenous biota on any of New Zealand's outlying islands. Only seven per cent – 6,800 hectares (26 square miles) – of the area of the islands is protected as conservation land, most importantly the 1,238 hectares (4.8 square miles) of the Tuku Nature Reserve, on the southern tablelands of Chatham Island, plus the Pitt Island Scenic Reserve and the two small island nature reserves of Rangatira and Mangere.

There is cause for future optimism, however. A renaissance in Moriori culture is leading to renewed interest by such islanders in traditional conservation measures such as *taiapure* (areas where fishing is carried out only in a customary Maori manner) and *rahui* (prohibition on access or harvesting). Another very positive development is the increasing use by islanders of Land Covenants to protect natural values on private lands. Most of all, the islanders' past suspicion of New Zealand Government agencies seems to be breaking down, at least in the field of conservation, reflecting the greater confidence displayed by the independent-thinking Chatham Islanders towards their future – and towards that of their remarkable natural and cultural heritage.

ABOVE Eastern (Buff) Weka (*Gallirallus australis hectori*), extinct on the mainland, has thrived on the Chatham Islands since introduction in 1905.

ABOVE The Pukeko (*Porphyrio porphyrio melanotus*) is now common on farmlands, especially those with Raupo swamps.

ABOVE The New Zealand Shore Plover (*Thinornis novaeseelandiae*), a small endemic wader that now survives in the wild only on Rangatira, was once widespread throughout all the main islands of New Zealand but has been almost wiped out by predation and breeding-habitat interference: only about 120 remain on this isolated island.

ABOVE The endemic Chatham Island Black Robin (*Petroica traversi*), one of the world's rarest birds. The story of its rescue from the brink of extinction is now part of the folklore of wildlife conservation. From a lowpoint of just five birds in 1980, the population has rapidly increased to nearly 150, most of them on Rangatira.

FAR LEFT The very rare Chatham Island Yellow-crowned Parakeet (Forbes' Parakeet; *Cyanoramphus auriceps forbesi*), a subspecies endemic to Mangere and Little Mangere islands, in the Chathams. It differs from the mainland Yellow-crowned Parakeet in being larger and having more brightly coloured plumage.

ABOVE Off Pitt Island are Mangere Island (right), Little Mangere Island (centre) and The Castle (left), three small, inaccessible and significant sanctuary islands. Mangere is a very important habitat for the endemic Chatham Island Yellow-crowned Parakeet, or Forbes' Parakeet.

LEFT The Chatham Island Forget-me-not (*Myosotidium hortensia*), a spectacular megaherb, growing up to a metre (40 inches) in diameter and with glossy rhubarb-like leaves, is the only species of its endemic genus, part of the Borage family. Once these plants dominated the coastal landscapes of the Chathams, but they have been largely eliminated through grazing by introduced stock.

Major Conservation Areas

Nearly 30 per cent of the land area of New Zealand – approximately 78,000 square kilometres (30,200 square miles) – is protected as conservation land. Areas of high nature-conservation significance are given special protection under one or other of the National Parks Act, Reserves Act or Conservation Act, with a small but growing number of marine habitats being protected under the Marine Reserves Act. Virtually all indigenous wildlife is protected under the Wildlife Act. The strictest categories of protected areas are:

- Special Areas within National Parks
- Nature Reserves
- Sanctuary Areas
- Ecological Areas
- Wilderness Areas
- Wildlife Sanctuaries
- Marine Reserves

The following list briefly describes the most important of the larger protected areas, running roughly north to south. It does not include areas of less than 20 square kilometres (7¾ square miles); some of the most important islands for nature conservation are smaller than this, but many of their nature reserves and wildlife sanctuaries are covered in the area descriptions in the main body of this book.

KERMADEC ISLANDS

Kermadec Islands Nature Reserve, 33 square kilometres (12¾ square miles), an isolated group of volcanic islands; only truly subtropical part of New Zealand.

Kermadec Islands Marine Reserve, 7,480 square kilometres (2,900 square miles), the second largest protected area in New Zealand and the outstanding marine protected area.

NORTHLAND, AUCKLAND AND COROMANDEL

Waipoua Sanctuary Area, 91 square kilometres (35 square miles), part of the largest remnant of the great Kauri forests of Northland. The sanctuary contains several massive Kauri, including Tane Mahuta, one of the largest trees in the world.

Rangitoto Island Scenic Reserve, 23 square kilometres (9 square miles), the young, forest-covered, symmetrical lava cone at the entrance to Auckland's Waitemata Harbour.

Little Barrier Island Nature Reserve, 28 square kilometres (11 square miles), a heavily forested volcanic remnant in the outer Hauraki Gulf, and one of the most important island sanctuaries in New Zealand. Little Barrier is one of three islands managed as Kakapo habitats.

Coromandel Forest Park, 750 square kilometres (290 square miles), an area of old volcanic mountains and largely cutover Kauri forest on the Coromandel Peninsula. It contains important historic relics of the Kauri-logging era and, in the Moehau Ecological Area at the northern tip of peninsula, a region of high-altitude vegetation of botanical importance.

BAY OF PLENTY AND CENTRAL VOLCANIC PLATEAU

Kaimai Mamaku Forest Park, 397 square kilometres (154 square miles), straddling the watershed between the Bay of Plenty and Waikato. This is similar to Coromandel (q.v.) in its forests and its historical associations with mining.

Lake Okataina Scenic Reserve, 44 square kilometres (17 square miles), an area of diverse volcanic landscapes and podocarp/broadleaf forest in the catchment of Lake Okataina, the most natural of the Rotorua Lakes.

Waioeka Gorge Scenic Reserve, 187 square kilometres (72¼ square miles), one of the largest and most accessible forested scenic reserves in the country. It lies on either side of State Highway 2 between Opotiki and Gisborne, as the road winds its way through the Waioeka Gorge.

Pureora Forest Park, 803 square kilometres (310 square miles), largely volcanic uplands lying to the west of Lake Taupo and containing the botanically important Pureora Mountain and Waihaha Ecological Areas; significant forest bird habitat.

Tongariro National Park, 765 square kilometres (296 square miles), New Zealand's premier volcanic protected area. The first National Park in the country. It is a World Heritage Site in recognition of the universal significance of its natural and cultural features.

Kaimanawa Forest Park, 767 square kilometres (297 square miles), a complex of beech-forested ranges and tussock-grassland plateaux lying to the east of Tongariro National Park. Popular hunting area for introduced deer.

EAST COAST

Urewera National Park, 2,126 square kilometres (822 square miles), high and remote, the largest forest wilderness in the North Island. Lake Waikaremoana is the key scenic feature; important habitat for forest birds.

Whirinaki Forest Park, 550 square kilometres (213 square miles), the outstanding dense podocarp forest of the North Island, with particularly tall Kahikatea, Rimu, Miro, Matai and Totara; contiguous with Urewera National Park.

Raukumara Forest Park, 1,151 square kilometres (445 square miles), an extremely rugged park centred on the Raukumara Range near East Cape. The park contains the 397-square-kilometre (154-square-mile) Raukumara Wilderness Area, centred on the Motu River, the first Wild and Scenic River designated in New Zealand.

TARANAKI/WHANGANUI

Egmont National Park, 335 square kilometres (130 square miles), comprising the three main volcanic cones of Taranaki – the beautiful symmetrical cone of Mount Taranaki (Egmont) and the eroded remnants of Pouakai and Kaitake. It is a park of dense forest, very high rainfall, steep, clear streams with many waterfalls, and, in winter, steep icy alpine slopes.

Whanganui National Park, 742 square kilometres (287 square miles), a maze of rivers and streams dissecting the soft papa (mudstone) rocks of the Whanganui hill country. Centred on the forested catchment of the Whanganui River, popular for canoe and jetboat travel. The area is important to the Maori owners of the bed of the river.

HAWKE'S BAY/MANAWATU/WELLINGTON

Kaweka Forest Park, 670 square kilometres (259 square miles), comprising the greywacke Kaweka Range and the open tussocklands and beech forests around the Ngaururoro and Mohaka Rivers. Popular hunting area, especially for Sika Deer.

Ruahine Forest Park, 936 square kilometres (362 square miles), a long, thin park straddling the Ruahine Range which separates Hawke's Bay from Rangitikei/Manawatu. Popular tramping and hunting area.

Tararua Forest Park, 1,172 square kilometres (453 square miles), one of the most rugged forested ranges (Tararua) in the North Island. Windswept and forested, with extensive subalpine tops, it is the major tramping area for the people of Wellington, Manawatu and Wairarapa.

NELSON/MARLBOROUGH

Mount Stokes Scenic Reserve, 44 square kilometres (17 square miles), the upper slopes of Mount Stokes, the highest point in the Marlborough Sounds. The summit provides outstanding views and is the only subalpine habitat in the sounds.

D'Urville Island Scenic Reserve, 41 square kilometres (16 square miles), straddles the central part of D'Urville Island, the largest island in the Marlborough Sounds. The island is possum-free, allowing the growth of some rare palatable plants such as Mistletoe.

Mount Richmond Forest Park, 1,884 square kilometres (729 square miles), a large mountainous park centred on the Bryant and Richmond ranges, which separate Nelson from Marlborough. Contains the Pelorus River and the mineralized landscapes of the Wairau Red Hills.

Abel Tasman National Park, 225 square kilometres (87 square miles), the smallest of New Zealand's 13 National Parks but a scenic gem of golden granite beaches and upland karst landscapes.

Kahurangi National Park, approximately 5,000 square kilometres (2,000 square miles), the newest and second-largest of New Zealand's National Parks. The park is centred on the Tasman Mountains, the core of which is left undeveloped as the 870-square-kilometre (337-square-mile) Tasman Wilderness Area. A huge park of remarkable geological and biological diversity. APPROVED BUT NOT YET GAZETTED

Nelson Lakes National Park, 1,018 square kilometres (394 square miles), a park of beech forests and greywacke mountains centred on the Spenser Mountains and the ranges around Lakes Rotoiti and Rotoroa. Contains the headwaters of the main rivers of the northern South Island – Buller, Wairau, Clarence, Waiau and Maruia.

Farewell Spit Nature Reserve, 114 square kilometres (44 square miles), an outstanding coastal landform and habitat for coastal wading birds; at the extreme north-west tip of the South Island.

Tapuae-o-Uenuku Scenic Reserve, 22 square kilometres (8½ square miles), covers the summits of the main peaks of the Inland Kaikoura Range. This remote range is a prominent landmark when viewed from Wellington across Cook Strait. It is the highest mountain range outside the Southern Alps and Fiordland.

CANTERBURY

Lake Sumner Forest Park, 1,058 square kilometres (409 square miles), a popular tramping area which includes the beech forest and alpine tops around Lake Sumner in northern Canterbury.

Lewis Pass National Reserve, 183 square kilometres (71 square miles), lies on both side of Lewis Pass (State Highway 7) at the southern end of Nelson Lakes National Park. A very scenic drive through large-stature beech forests.

Arthur's Pass National Park, 992 square kilometres (384 square miles), an alpine park containing the headwaters of the Waimakariri River (in the east) and the Taramakau River (in the west). Arthur's Pass was the first alpine pass on the main divide to be crossed by road (today's State Highway 73).

Mount Cook National Park, 699 square kilometres (270 square miles), the great alpine park of New Zealand. It contains all the 3,000-metre (9,800-foot) peaks of the Southern Alps (except for Mount Aspiring) and the Tasman Glacier, at 28 kilometres (17½ miles) the longest in the country.

WEST COAST

Buller Gorge Scenic Reserves (Upper and Lower), 119 square kilometres (46 square miles), lie along the beautiful Buller River (and State Highway 6) between Murchison and Westport. The landscape is a blend of wild river, beech forests, earthquake features and mountain gorges.

Victoria Forest Park, 2,107 square kilometres (815 square miles), the largest of the forest parks. It is centred on the Victoria and Brunner Ranges but includes also large areas of valley-floor beech forests, especially of the Maruia, upper Inangahua and upper Grey Rivers. The park is rich in historical gold-mining sites.

Paparoa National Park, 306 square kilometres (118 square miles), lies between the jagged crest of the Paparoa Range and the Tasman Sea. It has a spectacular coastline of limestone cliffs, many caves and other karst features, and the famous 'Pancake Rocks' of Dolomite Point near Punakaiki.

Westland National Park, 1,176 square kilometres (455 square miles), New Zealand's great 'mountains to the sea' park. It contains both the Franz Josef and Fox glaciers, and a wonderful range of lakes and other landforms of glacial origin. The lowland podocarp forests of Okarito and Waikukupa are some of the most impressive in the country.

Hooker/Landsborough Wilderness Area, 410 square kilometres (159 square miles), lies around Mount Hooker in the upper reaches of the wild and remote Landsborough River in South Westland.

OTAGO

Mount Aspiring National Park, 3,555 square kilometres (1,375 square miles), the third largest of New Zealand's National Parks. It has an alpine wilderness core (the Olivine Wilderness Area) and a wide range of forests – from the beech forests of the eastern valleys to the swamp podocarp forests of the Haast coastal plain in the west. Mount Aspiring (3,027m) is the most famous landmark in the park.

Catlins Forest Park, 606 square kilometres (234 square

miles), a remote park based on the impressive coastline of south-east Otago. It has luxuriant podocarp/hardwood forests – including the 92-square-kilometre (35½-square-mile) Ajax Ecological Area – and plentiful coastal wildlife, including penguins, seals and seabirds.

SOUTHLAND

Fiordland National Park, 12,570 square kilometres (4,860 square miles), a vast wilderness landscape, by far the largest park in New Zealand and one of the world's great wild National Parks. Notable features are its fiords (such as Milford and Doubtful sounds), its role in the survival of the Takahe and the Kakapo, its large number of endemic alpine plants and its hundreds of lakes (of which Te Anau and Manapouri are best-known). It also includes several of New Zealand's Great Walks – the Milford, Routeburn, Hollyford and Kepler Tracks.

Waituna Wetlands Scientific Reserve, 36 square kilometres (14 square miles), a coastal wetland lying close to the city of Invercargill. It is unusual in that it contains a number of plants normally found only in subalpine areas.

SOUTH-WEST NEW ZEALAND

Te Wahipounamu (South-West New Zealand) World Heritage Area, 26,000 square kilometres (10,055 square miles), encompassing virtually all of the south-west of the South Island – 10% of the area of New Zealand. It includes Mount Cook, Westland, Mount Aspiring and Fiordland National Parks, as well as the conservation land contiguous with these parks (including the lowland swamp podocarp forests of South Westland). It is New Zealand's outstanding contribution to the protection of the world's natural heritage, especially with its Gondwanan flora and fauna and its expression of past and present glaciation.

STEWART ISLAND

Most of Stewart Island – 1,722 square kilometres (666 square miles) – is conservation land, except for some significant blocks of Maori land. However, two large protected areas deserve special mention:

Anglem Nature Reserve, 170 square kilometres (66 square miles), protecting the subalpine plant communities in the northern part of the island around Mount Anglem – at 980 metres (3,215 feet) the highest point on the island).

Pegasus Nature Reserve, 674 square kilometres (261 square miles), covering almost the entire southern wilderness of the island. It includes the impressive granite domes of the Tin Range and the Fraser Peaks and the forests around the intricate waterways of Port Pegasus. This reserve was the last known large-island habitat of the Kakapo (now all transferred to smaller, predator-free islands).

SUBANTARCTIC ISLANDS

Auckland Islands Nature Reserve, 626 square kilometres (242 square miles), comprising four main islands: Auckland, Adams, Disappointment and Enderby. They are by far the largest and probably the most botanically important of New Zealand's subantarctic islands. Although there are introduced mammals on all except Adams Island, the islands as a group have outstanding value as habitats for marine mammals and seabirds.

Campbell Islands Nature Reserve, 113 square kilometres (43¾ square miles), the most southerly of New Zealand's protected areas (52 degrees 30 minutes South). Much of the island's flora was modified by introduced grazing animals, but it has recovered rapidly since their removal. The seabirds of the main Campbell Island have been seriously affected by introduced Norway Rats and cats, but the island group is still an exceptional breeding habitat for Albatross.

Bibliography

The books listed in this bibliography deal primarily with New Zealand's natural history and conservation at a national level. It should be noted that there are also numerous regional publications on New Zealand's natural landscapes and wildlife. The Department of Conservation publishes handbooks to most of the National Parks, and an excellent series of special maps covers all the National Parks and major protected areas; these 'Infomaps' are published by the Department of Survey and Land Information. Finally, the Department of Conservation also produces a comprehensive range of brochures on local natural areas.

TRAVEL

Allan, E. (1989) *AA Guide to New Zealand*. Weldon, Sydney.

DuFresne, J. (1982) *Tramping in New Zealand*. Lonely Planet, South Yarra.

McLauchan, G., Lawrence, M. (1987) *Insight Guide to New Zealand*. APA Productions, Singapore.

Pickering, M., Smith, R. (1986) *75 Great Tramps for New Zealanders*. Reed Methuen, Auckland.

Reader's Digest Services. (1981) *Wild New Zealand*. Reader's Digest, Sydney.

Williams, J. (1994) *New Zealand: A Travel Survival Kit*. Lonely Planet, Hawthorne.

BIOGEOGRAPHY AND ECOLOGY

Barnett, S. (1985) *New Zealand in the Wild: An Illustrated A-Z of Native and Introduced Birds, Mammals, Reptiles and Amphibians*. Collins, Auckland.

Bellamy, D., Springett, B., Hayden, P. (1990) *Moa's Ark: The Voyage of New Zealand*. Viking, Auckland.

Bishop, N., Gaskin, C. (1992) *Natural History of New Zealand*. Hodder & Stoughton, Auckland.

Brockie, B. (1992) *A Living New Zealand Forest*. David Bateman, Auckland.

Enting, B., Molloy, L. (1982) *The Ancient Islands: New Zealand's Natural Environments*. Port Nicholson Press, Wellington.

Kuschel, G. (1975) *Biogeography and Ecology in New Zealand*. Junk, The Hague.

Stephenson, G. (1986) *Wetlands: Discovering New Zealand's Shy Places*. Government Printer, Wellington.

GEOLOGY AND LANDFORMS

Cox, G.J. (1989) *Slumbering Giants: Volcanoes and Thermal Regions of the Central North Island*. Collins, Auckland.

Gage, M. (1980) *Legends in the Rocks*. Whitcoulls, Christchurch.

Homer, L., Molloy, L. (1988) *The Fold of the Land: New Zealand's National Parks from the Air*. Department of Scientific and Industrial Research, Wellington.

McCaskill, L.W. (1973) *Hold This Land: A History of Soil Conservation in New Zealand*. Reed, Wellington.

Molloy, L. (1988) *Soils in the New Zealand Landscape: The Living Mantle*. Mallinson Rendel, Wellington.

Soons, J.M., Selby M.J. (1982) *Landforms of New Zealand*. Longman Paul, Auckland.

Stevens, G.R. (1988) *Prehistoric New Zealand*. Heinemann, Auckland.

Stevens, G.R. (1980) *New Zealand Adrift: The Theory of Continental Drift in a New Zealand Setting*. Reed, Wellington.

Stevens, G.R. (1985) *Lands in Collision: Discovering New Zealand's Past Geography*. Information Series No. 161, Department of Scientific & Industrial Research, Wellington.

Thornton, J. (1985) *Field Guide to New Zealand Geology*. Reed Methuen, Auckland.

Williams, K. (1986) *Volcanoes of the South Wind*. Tongariro Natural History Society, Wellington.

VEGETATION

Brownsey, P.J. (1989) *New Zealand Ferns and Allied Plants*. David Bateman, Auckland.

Cockayne, L. (1967; 4th edn, ed. Godley, E.J.) *New Zealand Plants and their Story*. Government Printer, Wellington.

Crowe, A. (1990) *Native Edible Plants of New Zealand*. Hodder & Stoughton, Auckland.

Crowe, A., O'Flaherty B. (1992) *Which Native Tree? A Simple Guide to the Identification of New Zealand Native Trees*. Viking Pacific, Auckland.

Dawson, J. (1988) *Forest Vines to Snow Tussocks: The Story of New Zealand Plants*. Victoria University Press, Wellington.

Eagle, A. (1986) *Eagle's Trees and Shrubs of New Zealand*. Collins, Auckland.

Evans, A. (1987) *New Zealand in Flower: An Illustrated Guide to Native Flowering Plants*. Reed Methuen, Auckland.

Johns, J.H., Molloy, B.P.H. (1983) *Native Orchids of New Zealand*. Reed, Wellington.

Mark, A.F., Adams, N.M. (1986) *New Zealand Alpine Plants*. Reed Methuen, Auckland.

Salmon, J.T. (1992) *A Field Guide to the Alpine Plants of New Zealand*. Godwit, Auckland.

Salmon, J.T. (1991) *Native New Zealand Flowering Plants*. Reed, Auckland.

Salmon, J.T. (1980) *The Native Trees of New Zealand*. Reed, Wellington.

Wardle, J.A. (1984) *The New Zealand Beeches: Ecology, Utilization and Management*. New Zealand Forest Service, Wellington.

Wilson, H., Galloway, T. (1993) *Small-leaved Shrubs of New Zealand*. Manuka Press, Christchurch.

BIRDS

Butler, D. (1989) *Quest for the Kakapo*. Heinemann Reed, Auckland.

Ellis, B.A., Ellis, S.P. (1987) *The New Zealand Birdwatchers' Book*. Reed Methuen, Auckland.

Falla, R.A., Sibson, R.B., Turbott, E.G. (1985) *Collins Guide to the Birds of New Zealand and Outlying Islands*. Collins, Auckland.

Moon, G. (1992) *A Field Guide to New Zealand Birds*. Reed, Auckland.

Moon, G., Lockley, R.M. (1982) *New Zealand Birds: A Photographic Guide*. Heinemann, Auckland.

Peat, N. (1990) *The Incredible Kiwi*. TVNZ/Random Century, Auckland.

Reader's Digest Services (ed. Robertson, C.J.R.) (1985) *Reader's Digest Complete Book of New Zealand Birds*. Reader's Digest/Reed Methuen, Sydney and Auckland.

REPTILES AND AMPHIBIANS

Gill, B.G. (1986) *Collins Handguide to the Frogs and Reptiles of New Zealand*. Collins, Auckland.

Pickard, C.R., Towns, D.R. (1988) *Atlas of the Amphibians and Reptiles of New Zealand*. Science & Research, Department of Conservation, Wellington.

Robb, J. (1986) *New Zealand Reptiles and Amphibians in Colour*. Collins, Auckland.

INVERTEBRATES

Lessiter, M. (1989) *Butterflies and Moths*. Bush Press, Auckland.

Lessiter, M. (1990) *Interesting Insects*. Bush Press, Auckland.

Meads, M. (1990) *Forgotten Fauna*. Department of Scientific & Industrial Research, Wellington.

Meads, M. (1990) *The Weta Book: A Guide to the Identification of Wetas*. Department of Scientific & Industrial Research, Wellington.

Miller, D., Walker, A.K. (1984) *Common Insects in New Zealand*. Reed, Wellington.

MARINE AND FRESHWATER FAUNA AND FLORA

Dell, R.K., Heath, E. (1981) *Seashore Life*. Reed, Wellington.

Gunson, D. (1993) *A Guide to the New Zealand Seashore*. Viking Pacific, Auckland.

McDowell, R.M. (1990) *Freshwater Fishes: A Natural History and Guide*. Heinemann Reed/Ministry of Agriculture and Fisheries Publishing, Auckland and Wellington.

Morton, J., Cometti, R. (1985) *Margins of the Sea: Exploring New Zealand's Coastline*. Hodder & Stoughton, Auckland.

Morton, J., Miller, M.C. (1973) *The New Zealand Seashore*. Collins, London.

Paulin, C.D. (1992) *The Rockpool Fishes of New Zealand*. Museum of New Zealand, Wellington.

ISLANDS

Clark, M.R., Dingwall, P.R. (1985) *Conservation of Islands in the Southern Ocean*. IUCN, Gland.

Fraser, C. (1986) *Beyond the Roaring Forties: New Zealand's Subantarctic Islands*. Government Printing Office Publishing, Wellington.

Higham, T. (1991) *New Zealand's Subantarctic Islands: A Guidebook*. Department of Conservation,

Wellington.
Towns, D.R., Daugherty, C.H., Atkinson, I.A.E. (eds) (1990) *Ecological Restoration of New Zealand Islands: Papers Presented at the Conference on Ecological Restoration of New Zealand Islands.* Conservation Sciences Publication No. 2, Department of Conservation, Wellington.

MAORI IN AOTEAROA
Orbell, M., Moon, G. (1985) *The Natural World of the Maori.* Collins/Davis Bateman, Auckland.
Salmond, A. (1991) *The Two Worlds: First Meetings Between Maori and European, 1642-1772*, Viking, Auckland.

CONSERVATION MANAGEMENT
Major articles on nature conservation and management appear in two quarterly journals: *Forest and Bird*, published by the Royal Forest and Bird Protection Society, Wellington, and *New Zealand Geographic*, published by Academy Press, Auckland.

Butler, D. (1992) *The Black Robin: Saving the World's Most Endangered Bird.* Oxford University Press, Auckland.
Given, D.R. (1990) *Species Management and Recovery Plans for Threatened Plants: Botany Division Report.* Department of Scientific & Industrial Research, Christchurch.
Kelly, G.C., Park, G.N. (eds) (1986) *The New Zealand Protected Natural Areas Programme: A Scientific Focus.* Biological Resources Centre/Department of Scientific and Industrial Research, Wellington.
Molloy, J., Davis, A. (1992) *Setting Priorities for the Conservation of New Zealand's Threatened Plants and Animals.* Department of Conservation, Wellington.
Morris, R., Smith, H. (1988) *Wild South: Saving New Zealand's Endangered Birds.* TVNZ/Century Hutchinson, Auckland.
Norton, D.A. (ed) (1989) *Management of New Zealand's Natural Estate: Proceedings of a Symposium of the New Zealand Ecological Society held at the University of Otago 1988.* Occasional Publication No. 2, New Zealand Ecological Society, Christchurch.
O'Connor, K.F., Overmars, F.B., Ralston, M.M. (1990) *Land Evaluation for Nature Conservation: A Scientific Review Compiled for Application in New Zealand.* Conservation Sciences Publication No. 3, Centre for Resource Management, Lincoln University.
Wilson, C.M., Given, D.R. (1989) *Threatened Plants of New Zealand.* Information Series No. 166, Department of Scientific & Industrial Research, Wellington.

NATIONAL PARKS
Burton, R. (1987) *A Tramper's Guide to New Zealand's National Parks.* Reed Methuen, Auckland.
Thom, D. (1987) *Heritage: The Parks and the People.* Landsdowne Press, Auckland.

INTRODUCED PLANTS AND ANIMALS
Johns, J.H., MacGibbon, R.J. (1986) *Wild Animals in New Zealand.* Reed Methuen, Auckland.
King, C.M. (1990) *The Handbook of New Zealand Mammals.* Oxford University Press/The Mammal Society (NZ), Auckland.
King, C.M. (1984) *Immigrant Killers.* Oxford University Press, Auckland.

Glossary

Algae A large group of photosynthetic plants, lacking true stems, roots and leaves; almost all are aquatic. They include seaweeds and their freshwater allies.
Alluvial Adjective used of materials deposited from rivers.
Amphibians Animals (such as frogs) which are adapted to live either on land or in water.
Andesite Volcanic rock of chemical composition intermediate between 'basic' and 'acidic'.
Aotearoa The most frequently used Maori name for New Zealand.
Aquifer A rock layer that will absorb water or allow it to pass through.
Archipelago Closely grouped cluster of islands.
Arête Narrow, often jagged, mountain ridge, generally caused through two cirques eating into opposite sides of the ridge.
Argillite A dark, compact rock that formed from mudstone sediments that have been slightly metamorphosed (i.e., they have been hardened under pressure). Argillite is often found in bands with greywacke (q.v.).
Arthropods Animals with hard, jointed external skeletons and jointed limbs; they include insects, spiders and crustaceans (q.v.).
Axial ranges Series of mountain ranges forming the axis or 'backbone' of the land – in our context, of the North Island.
Basalt Igneous flow rock (usually a lava) of 'basic' composition (i.e., high in iron, manganese and magnesium) and lacking quartz.
Biota The plants and animals (flora and fauna – q.q.v.) of a given region.
Block mountains Mountains bounded on most sides by faults, so that they have a characteristic rectangular shape.
Brachiopod A bivalve mollusc with, on each side of the mouth, a long spiral arm used for procuring food.
Braided river River with a network of interconnecting convergent and divergent channels (resembling the strands of a braid of hair).
Calcareous Containing minerals of calcium (usually calcium carbonate).
Caldera Large, basin-shaped volcanic depression.
Cambrian One of the periods of geological time: 570-500 million years ago.
Cellulose The carbohydrate that is the chief structural component of the cell walls of plants.
Chronosequence Series of landforms (or soils on the landforms) whose formation can be read as a time sequence.
Cirque The amphitheatre-shaped head of a glaciated valley.
Conifers The group of trees that bear cones.
Continental shelf Bordering a continental mass, the relatively shallow belt that is below sea level, usually to a water depth of 200 metres (660 feet).
Cretaceous One of the periods of geological time: 135-65 million years ago. The Cretaceous is sometimes termed the Age of the Dinosaurs; the dinosaurs were indeed extinguished at its end, but their dominance in fact began about 200 million years ago, early in the Triassic period.
Crustaceans A class of arthropods (q.v.) including crabs, lobsters, shrimps, etc.
Diorite A coarse-grained igneous rock of intermediate composition (in terms of its content of silica and iron/magnesium minerals).
Diurnal Active in daytime.
Dyke A sheet of lava intruded into layers of other rocks.
Ecological niche A space within an ecosystem (q.v.). If an ecological niche is unoccupied within a particular ecosystem, an organism from outside may occupy it and thereby colonize the community.
Ecosystem A natural system formed by the interaction of organisms with their environment and with each other.
Endemic Restricted to a certain region, or to part of a region.
Eocene One of the epochs of the Tertiary (q.v.): 53-37 million years ago.
Epiphyte A plant which grows on the surface of other plants.
Exclusive Economic Zone A 200-mile (322km) band of sea extending out from the coast of all New Zealand territory, within which the nation claims exclusive economic rights, especially fishing.
Fauna The animal population of a given area.
Fell-field The upper part of the alpine zone, generally consisting of small scattered herbs and tussocks.
Fiord A long, narrow inlet of the sea bounded by steep mountain slopes. Fiords were formed when the sea flooded back into glacier-cut valleys as sea-level rose after the Pleistocene (q.v.) ice age.
Flora The plant population of a given area.
Foehn wind A warm, relatively dry, wind that descends (often with considerable strength) on the leeward side of a mountain range. Strictly, this term should be confined to the European Alps; in New Zealand these winds are generally termed 'nor'-westers'.
Frost-heave The raising of the soil surface through the expansion of ice in the underlying soil.
Fumarole Small vent in a volcanic area, usually giving rise to steam or gases under pressure.
Gendarme Rocky pinnacle on a narrow mountain ridge.
Glacier An enduring mass of ice: an ice-sheet or icecap, or a river of ice descending from a snowfield or mountain.
Gneiss A coarsely crystalline rock with alternating mica and granite-like layers. Gneisses form through the metamorphosis (*see* metamorphic rock) of either igneous or sedimentary rock.
Gondwana Often called Gondwanaland, the ancient supercontinent that existed in the southern hemisphere up to Triassic-Jurassic (q.q.v.) times. It contained precursors of modern South America, Africa, Madagascar, India, Australia, Antarctica and New Zealand.
Granite A granular, crystalline rock of igneous origin, consisting essentially of quartz, feldspar and mica. It is of 'acidic' composition (i.e., high in silica).
Great Walks Term used by the New Zealand Department of Conservation to describe New Zealand's higher-standard back-country walking tracks (e.g., Milford Track).
Greywacke Sand-sized grains of the characteristic sediments deposited in ancient oceanic basins, subsequently hardened into greyish rock. This rock contains quartz and feldspar as well as minerals of volcanic origin.
Habitat The environment in which a particular plant or animal lives.
Hanging valley Higher side-branch, usually of glacial origin, of a mountainous river system, normally connected to the main system by a waterfall or series of rapids.
Herb-field Lower part of the alpine zone, a dense community of alpine herbs with, usually, shrubs and tussock grasses.
Humus Decaying organic material in soil.
Igneous Adjective referring to rocks (usually crystalline) formed by the cooling of magma (q.v.) at or beneath the Earth's surface.
Ignimbrite Glassy rock, found as sheets, formed by the welding together of extremely hot particles of rhyolitic (q.v.) ash during volcanic eruptions. The name means 'fire rock'.
Indigenous Native to a particular region or country.
Inflorescence The collective flowers of a plant, often arranged on a central stem.
Intermontane basins Basins lying between, or surrounded by, mountain ranges.
Invertebrates Animals without a backbone (e.g., insects, worms, snails).
Isthmus Narrow piece of land separating two bodies of water, and connecting two larger pieces of land.
Iwi Maori tribe.
Jurassic One of the periods of geological time: 190-135 million years ago.
Karst A landscape created by water dissolving limestone, and characterized by widespread surface depressions, extensive subterranean caves and the virtual absence of surface drainage.
Kumara A sweet potato (*Ipomoea batatas*) of tropical origin, the major cultivated crop of pre-European Maori society.
Lahar A mud-flow composed mainly of volcanic ash lubricated by water from rapid snow-melt, torrential rain or a burst crater lake.
Lek A courtship behaviour in which male birds try to attract potential female mates to their courtship

areas by a combination of calls and displays.

Levee Elevated bank flanking the channel of a river.

Liana *or* **liane** A climbing plant with long, twining stems, typical of tropical or subtropical rainforest.

Lignin A naturally occurring organic material of complex chemical structure found as a 'cement' in the cell walls of woody plants.

Limestone A sedimentary rock, usually consisting of calcium and magnesium carbonates.

Loess Accumulated wind-blown dust.

Longshore drift The movement of sand and gravels along a shoreline through the effect of waves breaking obliquely on the beach.

Magma Molten rock from the Earth's mantle (the region beneath the crust), highly gaseous and mobile in nature and capable of escaping to the Earth's surface regions either through volcanism as a flow of lava or by being intruded into crustal rocks.

Mana whenua The spiritual power and prestige felt by Maori people through their link with the land (and, through it, with Papatuanuku, the Earth Mother).

Marine terrace A wave-cut platform later uplifted above sea level.

Mesozoic The 'middle' era (or era of 'middle life') of geological time: about 225-65 million years ago. The Mesozoic includes the Triassic, Jurassic and Cretaceous periods.

Metamorphic rock Rock that has been altered from its original state through intense pressure and/or heat, such as that generated through mountain uplift or volcanic activity.

Micas A group of crystalline silicate minerals. They are shiny and form very thin flakes that can easily be separated.

Miocene One of the epochs of the Tertiary (q.v.): 24-5 million years ago.

Monoculture Vegetation community consisting almost exclusively of a single species, or a single dominant species.

Montane zone The middle altitudinal belt, lying above the lowlands but below the subalpine region.

Moraine Poorly sorted, angular pieces of rock and debris carried by a glacier and deposited on the margins (lateral moraine), between tongues (medial moraine) or at the lower end (terminal moraine) of the ice.

Mustelids Members of the Carnivora family Mustelidae; in the New Zealand context, the term usually refers to stoats, weasels and ferrets.

Névé Freshly deposited snow in the accumulation region of a glacier.

Oligocene One of the epochs of the Tertiary (q.v.): 37-24 million years ago.

Orthoptera The animal order of insects.

Pakeha The name used by Maoris since early in the 19th century to refer to Europeans.

Palaeozoic The era of 'ancient life', about 600-225 million years ago; the first of the three main eras of geological time that have succeeded the long, misleadingly termed Azoic (era of 'no life'), or Precambrian.

Paleocene One of the epochs of the Tertiary (q.v.): 65-53 million years ago.

Parasite Organism living on, or in, another organism without being of benefit to its host.

Peat Partly decomposed mass of vegetation usually formed in a shallow wetland.

Peneplain A landform of low relief formed by prolonged erosion.

Peridotite A dark green, coarsely crystalline ultramafic (q.v.) rock.

Periglacial Adjective describing the climate, processes and features created by freeze-thaw action in a zone bordering ice-sheets.

Permian One of the periods of geological time: 300-225 million years ago.

Pleistocene The first epoch of the Quaternary period of geological time, which succeeded the Tertiary (q.v.): 2 million to 10,000 years ago. The Pleistocene was marked by the most recent ice age.

Pliocene One of the epochs of the Tertiary (q.v.): 5-2 million years ago.

Podocarps The group of southern conifers (q.v.) including the genera *Dacrydium*, *Podocarpus* and *Dacrycarpus*. The name means 'seed with a foot', referring to the coloured, fleshy 'fruit-like' stalk on the end of the seed.

Pyroclastic 'Broken by fire'; adjective describing the material ejected in various forms during a volcanic eruption. Sometimes referred to as tephra (q.v.), this includes volcanic ash, scoria, pumice, lapilli, lava, ignimbrite, etc.

Quartz Crystalline silica, a hard glassy-looking mineral.

Rain-shadow region An area with a relatively low annual precipitation because sheltered by a mountain range from the prevailing rain-bearing winds.

Rhyolitic Referring to volcanic rock rich in silica and low in iron, manganese and magnesium. The viscosity of rhyolitic lava leads to violent volcanic eruptions; the lava is often ejected full of gas bubbles, and the resultant light pale rock is termed 'pumice'.

Rift valley Linear depression of the Earth's crust between two parallel faults.

Ring plain Low-angle plain surrounding a typical basaltic or andesitic volcano, built up by the successive lava flows and lahar.

Saltpan Impermeable layer in a soil, formed through the deposition of salt crystals.

Scarp Steep slope terminating in a plateau.

Schist A type of metamorphic rock (q.v.). Various forms of mica (q.v.) appear as characteristic silver flecks in the rock.

Scoria Dark, cindery fragments of lava ejected from a volcano.

Scree slope *or* **talus slope** Sloping pile of loose rock fragments eroded from a cliff or mountain face and accumulated at the base of the cliff.

Serpentine Ultramafic (q.v.) metamorphic rocks (q.v.); they are greenish with a greasy appearance.

Siltstone A fine-grained rock consolidated from silt.

Sinter Deposit of siliceous minerals around the mouth of a geyser or hot spring.

Stack Isolated rock pillar rising steeply from the sea.

String bog Type of bog formed in a periglacial (q.v.) environment, consisting of levees (q.v.) of peat and cushion plants interspersed with depressions that are often filled with water to form shallow tarns (q.v.).

Tableland Undulating area of high relief.

Talus slope Synonym for scree slope (q.v.).

Taonga To Maori, all treasures, both tangible and intangible, including those inherited from the past, for the present, and for future generations.

Tapu The Maori concept of sacredness, either of persons or of objects.

Tarn Small lake, usually in an alpine environment and often of glacial origin.

Taxon (plural **taxa**) Any named group of organisms, whether at subspecies, species, genus, family or higher level.

Te Papa Atawhai Maori name for the New Zealand Department of Conservation; it can be translated as 'those who care for the special places which contain the nation's treasures'.

Tephra Collective term for all material, regardless of size, ejected through the air by a volcano. Compare 'pyroclastic' (q.v.).

Tertiary One of the periods of geological time: 65-2 million years ago. The Tertiary was marked by the rise of the mammals (among animals) and of the flowering plants. It was composed of the following epochs: Paleocene, Eocene, Oligocene, Miocene and Pliocene (q.q.v.).

Tiller Those shoots of a plant that continue to spring from the bottom of the original stem.

Tohunga Maori priest, or person steeped in traditional knowledge.

Tor A mass of residual rock capping a hill after exposure by erosion.

Track-and-bowl systems An intricate network of tracks and arenas formed by male Kakapo to facilitate their lek courtship behaviour; the display arenas are generally bowl-shaped and are often backed by large rocks which reflect the birds' 'booming' calls.

Treaty of Waitangi Treaty signed in 1840 between Maori chiefs and the representatives of the British Crown, whereby the Maori people embraced UK sovereignty in return for the guaranteeing of many of their traditional rights.

Trilobite A fossil arthropod characteristic of rocks from the lower Paleozoic (q.v.) era.

Tundra Extensive treeless plain, usually covered with bare ground or a low vegetation of mosses, lichens or dwarf shrubs.

Ultramafic Referring to igneous rocks consisting of minerals with a high level of iron and magnesium and low levels of silica.

UNESCO United Nations Educational, Scientific, and Cultural Organization.

Whakapapa Maori genealogy, traced back through ancestors to the primal parents Ranginui and Papatuanuku. Places having a historical association with an important ancestor are an important link to the natural world.

Windthrow The uprooting of (generally shallow-rooted) trees by wind-storms, usually with the trunks being snapped. Large areas of forest can be windthrown, the fallens trees lying oriented in line with the direction of the wind.

Index

Page numbers in **bold** *type refer to captions; those preceded by* m *refer to maps.*